# LIFE AT THE CENTER OF THE ENERGY CRISIS

**A Technologist's Search for a Black Swan**

# LIFE AT THE CENTER OF THE ENERGY CRISIS

## A Technologist's Search for a Black Swan


**George H Miley**

University of Illinois at Urbana-Champaign, USA

NEW JERSEY • LONDON • SINGAPORE • BEIJING • SHANGHAI • HONG KONG • TAIPEI • CHENNAI

*Published by*

World Scientific Publishing Co. Pte. Ltd.
5 Toh Tuck Link, Singapore 596224
*USA office:* 27 Warren Street, Suite 401-402, Hackensack, NJ 07601
*UK office:* 57 Shelton Street, Covent Garden, London WC2H 9HE

**Library of Congress Cataloging-in-Publication Data**
Miley, George H. (George Hunter), 1933–
Life at the center of the energy crisis : a technologist's search for a black swan / George H. Miley, University of Illinois at Urbana-Champaign, USA.
pages cm
Includes bibliographical references.
ISBN 978-9814436489 (hardcover : alk. paper)
1. Nuclear engineers--United States--Biography. 2. Nuclear energy--Research--United States--History. I. Title.
TK9014.M55A3 2013
621.48092--dc23
[B]

2012046467

**British Library Cataloguing-in-Publication Data**
A catalogue record for this book is available from the British Library.



Typeset by Stallion Press
Email: enquiries@stallionpress.com

**Printed in Singapore by B & Jo Enterprise Pte Ltd**

# Contents

# Dedication

My wife Liz and I were married in 1958 in Ann Arbor, MI, one day after my final PhD thesis defense. My friends considered that "daring" — what if I didn't pass? But I did, and much has happened since then. In this book I describe my teaching and technology endeavors, and attempt to explain the reasoning behind my personal search for a "Black Swan," a discovery that would be a game-changer in energy sources and their use.

As our friends know, I have traveled to many professional meetings throughout my career, and some say that traveling to these meetings has been one of my "hallmarks." In

the photo above and on the left, Liz's favorite flower pin is just visible on her left shoulder. She still wears a frog, bird, dragonfly, or leaf pin to many occasions. These signify one of her "hallmarks," namely her love for all things associated with nature. Her eyes are always searching to see the beauty in nature's creatures, and in the trees, flowers, sky, and water around us. As an update, one of my more recent photos is at the right.

This book is dedicated to my family, former students, colleagues, and friends who helped and often led me along the zigzag path while searching for the elusive "Black Swan." My wife, Liz, and our children, Susan and Hunter, were constant and understanding companions on this journey of many years. Without all of these people, I would have ended up in a quicksand bog long ago. Hunter and Susan are shown here with us at a family gathering in 2002. Hunter and his family live in Walnut Creek, CA, while Susan and her family live in Bloomington, MN. Thus family gatherings are not as frequent as we would like, but always something we look forward to.

## Acknowledgments

Many people were instrumental in the writing of this book. In particular, I would like to thank Robyn Bachar, for her able and dedicated assistance in editing and proofreading; Prajakti Joshi Shrestha, who was a great help in proofreading; and Alvin Chong, my editor, who helped bring everything together.

# Foreword

*Heinz Hora*
*Professor Emeritus, Department of Theoretical Physics*
*University of New South Wales, Sydney, Australia*

I am most pleased to comment on this fascinating book in which George Miley provides a brief history along with some intimate thoughts about his long career in science and technology. When Sir Isaac Newton was asked how he was able to discover key physics and mathematics behind our modern world, he replied that he had the opportunity to "stand on the shoulders of giants". This permits a comparison with George Miley. His remarkable achievements in science and engineering have moved on from existing technology to provide new visions for the coming century, and in the process he has inspired numerous students to begin on this path.

The diversification of groundbreaking discoveries by Miley has involved starkly different directions. However, there has always been the common thread of nuclear energy involved. Even his recent involvement with fuel cells for automobiles has a nuclear origin in that nuclear power plants are viewed as the energy source of choice for production of hydrogen needed for full-scale operation of the future "hydrogen economy." His work is generally characterized as attacking an important application or societal need, e.g., his pioneering work on the plasma physics of neutron sources using Inertial Electrostatic Confinement (IEC) fusion has most important implications for neutron activation analysis (NAA) used in Homeland Security detection systems. Miley's discoveries related to driven subcritical operation for improved safety in nuclear fission reactors seems even more important now in light of the nuclear reactor damage at the Fukushima Daiichi nuclear power plant causing radioactive leakage following the earthquake-tsunami in Japan. His other pioneering work in lasers, including the very first electron beam excited laser and the advances in using radiation from nuclear fission reactors to pump lasers (nuclear-pumped lasers, or NPLs) established Miley as a global authority in the field.

Also noteworthy is his work as editor of several major professional journals, his most successful and appreciated teaching and mentoring of students, and his splendid organizing ability. An example of the latter is the influential Laser-Plasmas Interaction Workshop series he coordinated for many years at the Naval Postgraduate School in Monterey, CA. This workshop latter became the current major international meeting series on Inertial Confinement Fusion Science and Applications of Inertial Confinement Fusion

(ICFSA). His effort as an Executive on the Board of Directors of SAFE (Society to Advance Fusion Energy) spearheaded their successful campaign to promote a large increase in the budget for nuclear fusion research during the presidency of Jimmy Carter.

In 1970 he began working out the first details for the nuclear fusion cross sections of p–$^{11}$B in the often cited book, *Fusion Cross Sections and Reactivities* (DOE Report C00–2218–17, Gov. Publishing Office, Springfield, VA, 1974). No neutrons are directly produced in this reaction, suggesting the possibility of a nuclear fusion reactor that would produce, on a per unit energy basis, less radioactivity in the entire process, including waste products, than from burning coal. In his profound 1976 book *Fusion Energy Conversion*, Miley elaborated the case for advanced fuel fusion energy well. Indeed, his fusion research has continuously been directed towards low neutron emission nuclear energy production, corresponding to the dream of a "golden age" where the ten million times more efficient nuclear energy replaces most of the chemical energy from fossil sources with extremely clean and low cost power. The challenge mankind faces in getting to this goal is reflected in Miley's view of "Patience and Optimism," the title of his lecture when receiving the Edward Teller medal in 1995.

The difficulty in plasma physics involved in fusion power was so well expressed by Edward Teller (*Memoirs*, Perseus Publishing, Cambridge MA, 2001) when referring to the situation in 1952. "Research on controlled fusion means dealing with the hydrodynamics of plasma. I have a thorough respect for the fearsome nature of hydrodynamics, where every little volume does its own thing. The same complications occur in planning a thermonuclear explosion. But an explosion occurs in such a short time that many of the complicated phenomena have no chance to develop. *I have no doubt that demonstrating controlled fusion will be even more difficult*" (p. 344). Miley has certainly also recognized this challenge, and has always strived to find ways to achieve fusion power in a reasonable time. Indeed, his work continues to help the community along this path.

This problem with plasma physics and hydrodynamics as related to controlled fusion power is still not fully solved some 60 years after Teller's comment, though much has been learned. Only research with great depth and breadth can lead us forward, as suggested by Lord May of Oxford (*Nature,* Vol. 238, 1972, p. 413; Vol. 469, 2011, p. 351) who ingeniously employed approaches used in theoretical physics to master complex systems mentioned by Teller. Thus Lord May proposed solutions based on asking "how will a large complex system be stable," and began applying this approach to population systems in zoology and finally to "systemic risk in banking systems." In a similar way George Miley has approached problems by a combination of theory, computations, and experiments, saying, "I cannot be categorized as either a pure theorist or pure experimentalist. My guiding light came from Enrico Fermi's ability to analyze complex physics with simple elegant treatments that avoid excessive math but always supplied just enough to solve the problem at hand."

Miley has frequently stepped into the forefront of new explorations. Indeed, Miley was aware in advance of the crucial disclosure by Stephen Jones (March 1989) that led to the decision for a public announcement of cold fusion by Fleischmann and Pons. It is a tragedy how the subsequent confusion by claims, disclaims, and some untruths has poisoned this whole development. This turned many people, including most of the physics community, against cold fusion. Characteristically, Miley has done his best to keep an open mind and to dig down to the truth with his own independent research and also, as editor of several journals, sending cold fusion papers on to open minded reviewers (many physics journal editors have rejected such papers outright without review). His experimental work is very different from most scientists who still work in cold fusion. He discovered the low energy nuclear reactions (LENR) that produce nuclei over the whole range of elements detected using very sensitive analytic equipment in the Frederick Seitz laboratory of the University of Illinois as documented in *Low Energy Nuclear Reactions and New Energy Technologies Sourcebook II* (J. Marwan *et al.*, eds., Am. Chem. Soc./Oxford University Press, 2009, pp. 235–252). Miley's measurements show a remarkably small local maximum of the element distribution near the nucleon number A = 154. In comparison such a local minimum occurs in the fission of Uranium at A = 119. This comparison confirms both to be a Maruhn-Greiner effect and provides a solid proof of the phenomenon of LENR, thus opening important directions for further exploration.

Another most intriguing discovery related to the LENR work is Miley's SQUID and superconducting measurement (*Phys. Rev.*, B72, 2005, 212507) of the generation of clusters with more than 100 deuterons imbedded in the volume of one of the void defects (Schottky defect) created in palladium. This is parallel to the later measurements of ultrahigh densities of clusters by Leif Holmlid in Gothenburg/Sweden where anomalously low laser intensity emissions of electrons and nuclear reactions have been detected (*Laser and Particle Beams*, Vol. 28, 2010, p. 317). Both effects have been explained in a theory involving a Bose-Einstein like condensation.

I am sure that readers of this book will enjoy and be most impressed with what Miley has achieved in over 50 years at the University of Illinois while working with students and associates and worldwide contacts.

*16 March 2011*

# Preface

## Why a "Black Swan"?

*Photo of my sighting of a White Swan family on the Thames River in Oxford during a visit to the UK Fusion Laboratory, Culham, in the 1970s. As explained in the text, we often mentally associate a White Swan with "good" and a Black Swan with "bad." However in this book, the sighting of a Black Swan signifies an unexpected momentous event or discovery that is hopefully good, but might have a black side too.*

What is a "Black Swan" and why should anyone be searching for one? I first heard the term "Black Swan" in 2009 when I attended a meeting in San Francisco on Inertial Confinement Fusion (ICF). John Nuckolls, former head of Lawrence Livermore National Laboratory (LLNL), presented a keynote talk about the "fusion ignition campaign" ongoing at the National Ignition Facility (NIF) at LLNL. At the meeting, John Nuckolls said the only thing that could stop them from achieving this historic milestone was a "Black Swan" event. After his talk, I asked Nuckolls the meaning of a "Black Swan." He said that the terminology came from a recent bestselling book by Nassim Taleb, titled *The Black Swan*. Simply put, a Black Swan is an unexpected event which has immense direction-changing implications. We all are intuitively aware of such events, usually giving them other names, such as "game-changing," "revolutionary," or "earth-shattering." However, Taleb provided a fresh look at such events and how they can affect us at all levels, including personally, nationally, and globally. As I considered this, I mentally equated a Black Swan to possible new physics phenomena which everyone had overlooked. Also, I thought to myself, "The NIF people want to avoid the Black Swan, while I spend my time trying to find one!" That is, most of my research has been aimed at finding new phenomena which would have a dramatic impact on new energy sources and their applications. Thus, my Black Swan event, I hoped, would be revolutionary with a good (not bad) impact on mankind.

As the present book neared completion in the summer of 2012, it now seems ironic that the NIF ignition did in fact hit a Black Swan in the form of unexpected behavior of the laser interaction with the ICF target ablator material. Thus the optimistic prediction of igniting a target in late 2011 did not occur. Department of Defense (DOD) funding for the ignition campaign is to end in late 2012, after which less time (i.e., "shots") on the NIF laser will be devoted to ignition studies while more fundamental high density physics and other DOD projects will gain time. Thus there was much pressure on NIF scientists to find a way around the Black Swan event and achieve ignition by the end of 2012. One approach is to change to new ablator materials that are hoped to behave more as predicted in the complex computation of laser implosions. This event, however, has been a classic one where NIF scientists had great confidence in their computer simulations using the "famous" LASNEX code (classified due to also being used in weapon design) prior to recent experiments that didn't follow LASNEX predictions. The big banks and Wall Street analysts had similar confidence in their investing abilities prior to the 2009 financial collapse.

In reading Taleb's book, I found that he ascribed important characteristics to a Black Swan. It is an event which lies outside of our normal statistical coverage in the field, i.e., an outlier. Such events characteristically occur when no one is expecting them, but they have a profound effect. Clearly such an event could have either good or bad consequences, or even a mixture of the two. *Black Swan* often carries a negative implication to most people, so it might seem that we should look for white swans. Taleb's choice of a Black Swan is based on the fact that they are much harder to find than white ones. I am sticking with that here and hope the reader will accept the association of good as well as bad with a Black Swan event. In his book, Taleb focuses on bad events, like the 2008 collapse of the financial markets in the United States that rapidly triggered a worldwide economic recession. He also cites other examples such as extreme weather conditions, religious events, and personal events such as meeting one's future wife. Indeed, Taleb says that if you think about your own experience, the key changes in direction come from Black Swan events in your own work or personal life, along with the uncontrollable external events like the 2008 financial crash.

What are examples of a good Black Swan event? Great inventions or discoveries starting from the wheel, the steam engine, nuclear energy, the computer, and the Internet are examples that immediately come to mind. Ultimately, all of these have also had their negative sides, but all have revolutionized people's lives, hopefully providing more positive benefit to mankind than negative. I am seeking such a beneficial Black Swan in energy. We desperately need a revolution in that area as a keystone for establishing a sustainable future society on earth.

Taleb says that another characteristic of the Black Swan event is that once it happens, most "experts" manage to explain it and act as if they anticipated its coming. Taleb says this is because many of these experts have convinced themselves that they know more than they actually do. He goes on to state that their ignorance of crucial insights needed to predict Black Swans is often hidden by a "cloud" of computations and equations. These analyses presumably explain everything, but, in fact, miss the key elements leading to a Black Swan. Still, as a result of Taleb's book, the financial world is much more attuned to the Black Swan concept now than the scientific world. For instance here is an interesting quote from the financial newsletter "The Complete Investor" dated August 20, 2012. "The trouble with Black Swan events is that they're sort of like Heisenberg's uncertainty principle: Once you start thinking about them, or listing what the potential Black Swan events is or could be, almost by definition they are no longer Black Swans. With that said, and fully aware that we are contradicting ourselves to at least some extent, here's one candidate for a Black Swan…"

For the technologist or scientist, this means that we cannot fully identify a roadmap of studies and experiments that will lead to one. All we can do is follow the traditional scientific approach of first defining the problem, analyzing it, and then seeking a solution. Then, we keep our fingers crossed. As they say, there is an element of luck involved in solving a truly new problem, along with hard work and a dab of inspiration. And you have to be in the right place at the right time. Looking back over my research, I must conclude that I have achieved finding a few "Mute" Swans, but I am still searching for the Black one. I originally wanted to call this consolation prize a Gray Swan, but could not find that species in my copy of *Field Guide to Birds of North America*. However, this guide identifies "Mute Swans" as an old world species which, after introduction here, are seen in parks and also in the wild. They have an orange bill with a black head marking, and their voice, "a low grunt," is seldom heard. The more I mulled this over, the more I came to realize that these characteristics were just what I wanted to represent the symbolism in the context of this book. Mute Swans, located in isolated areas, are more easily discovered than Black Swans, which remain an elusive enigma.

Patience and optimism are the essential elements needed to continue the search for a Black Swan, because the search can take years, or forever. I have tried to incorporate these characteristics into my life and studies to stay on course. However, the path is zigzagged, making it hard to follow. I believe I am still on the right course, but time will tell if I am.

A book about my varied research was first proposed to me by Tom Valone, President of the Integrity Research Institute in Washington, DC. My first concern was, "Why would anyone want to read about my work and struggles?" Tom did not mention "Black Swan" but still the connection came to me. I decided that some common ground might be found by readers who are also on a mission to find a "good" Black Swan. Their Black Swan may be different from mine, but it is no doubt just as important and elusive. Maybe some readers have succeeded in finding their Black Swans, but I suspect that many are in the same boat with me and may have to settle for a Mute Swan. In any case, perhaps we can enjoy thinking about each other's journey and in doing that refresh our patience and optimism.

The breadth of my search, i.e., the varied directions I have traveled, makes it difficult to explain my path. To help clarify the reasons for moving from one direction to another, I describe my zigzagged path in graphical form in the Appendix. As suggested by Tom Valone, I have also included a few representative publications from each "leg" (chapter) in the path, using papers that are to some extent readable, i.e., they contain less specialized technical and mathematical detail. A short collection of photographs is also included. Their selection was mainly based on which old photos I could easily find to provide some visual insight into the "flavor" of each "leg" of the journey.

When reading this book, or the references listed, you may no doubt wonder if any of these often futuristic-sounding concepts ever become practical. The jury remains out on many, but a few have already made an impact. For example, various direct nuclear-to-electrical conversion methods were later used in commercial radiation detectors and for beam energy recovery at large ion accelerator facilities. Also an Inertial Electrostatic Confinement (IEC) fusion device has been used for commercial neutron activation analysis in a commercial quality control program at their aerospace factory in Germany. Admittedly, these were not the applications originally intended, but such unanticipated spin-offs are the way research often works. Or, in terms of swans, even the discovery of a Mute Swan, while more likely than sighting the Black Swan, is still unpredictable.

As I was writing this book, several events occurred that deserve mention. On March 3, 2011, Charlie Rose interviewed Nassim Taleb on his PBS TV show. Taleb's book, *The Black Swan*, has sold over 3 million copies and is listed by the *New York Times* among the top twelve most influential books of the decade. Nassim said that he has been reflecting on the book and concluded that the world has become so complex that we cannot possibly anticipate the next Black Swan event. Meanwhile this complexity can also lead to dire consequences. Hence he concludes that rather than trying so hard to predict, we must prepare by having a reserve or cushion to fall back on. Nassim further said that unfortunately our government officials don't seem to catch on. The present deficit gives us virtually no cushion for another Black Swan financial event. He feels it is essential that we "bite the bullet" and build a cushion. I don't want to get into politics here, but relative to the present book and my career, I think that I instinctively tried to build in a cushion. That is one way of looking at the diverse, zigzag path discussed later. If one direction didn't work, I wanted another one ready to fall back on!

As I neared completion of this book, I attended the inaugural Summit Meeting of the new government agency APRA-E. This agency was founded in 2009 with the mission of fostering revolutionary (disruptive) inventions in the clean energy area that will "solve the U.S. energy needs, control energy effects on climate change, create new jobs, and regain technological leadership for the U.S." Thus, we now have a government agency with the mission of finding Black Swans in the energy area. I never anticipated that our government would be so bold. Here we have a whole government agency expecting to do quickly what I have devoted years to without success (only "near misses"). Maybe ARPA-E too will, in reality, be happy with a few "Mute Swans"? (I cannot possibly relate here all that was said about ARPA-E, but the interested reader should read the National Academy Report, "The Gathering Storm," from the Augustine Committee, which originally proposed formation of ARPA-E.)

There are a few points about ARPA-E that are particularly relevant to my Black Swan search. It was said that unless a fraction of the ARPA-E projects fail, APRA-E managers have not taken on sufficient risk. Well, in my personal search for the Black Swan, described here, I did not think that way. I didn't shy away from what I viewed as lofty goals, but failure was not a part of my thought process either. Failure for an individual is demoralizing, even if we are determined to learn from our mistakes. Thus, what I did, in retrospect, was redefine my goals as I hit roadblocks, trying to find something obtainable to be proud of. Thus, if the Black Swan stayed hidden, I sought a Mute Swan. This attitude fits, I believe, with the often cited story about Thomas Edison, who responded to a news reporter who asked if he was depressed about the failures he had suffered up to that point in development of a light bulb filament by saying: "If I find 10,000 ways something won't work, I haven't failed. I am not discouraged, because every wrong attempt discarded is another step forward." As such, his disintegrating filaments were setbacks along the route to the light bulb, not failures in the sense of ARPA-E project failures. Failures in the latter sense imply a fatal flaw, not just a setback. Unfortunately, we may become impatient and confuse a setback with a fatal flaw, but there is a tremendous difference. If it is a setback, we should renew our determination and forge ahead. If a fatal flaw is found, one should stop beating their head against the wall and change directions. However, often the researcher does not have all of the data needed to easily decide between the two. Making the distinction represents one of the many possible pitfalls along the zigzag path to the Black Swan.

Another event that raised questions important to this book is the movie, *Black Swan*, released during the first week of December 2010. The movie stars Natalie Portman playing the ballerina, Nina. Natalie received an Academy Award for the performance (talk about a Black Swan event for her!). A review by Susan Wloszczyna in the December 3, 2010, issue of *USA TODAY* stated that Nina is a, "driven ballerina who precariously pirouettes between art and madness in the delirious psychosexual thriller." Nina was searching for her Black Swan of stardom, and the process became so focused that she lost her balance of life. That is something that all of us searching for a Black Swan must avoid — we must learn to keep laughing and enjoying life as we zigzag along the path searching for our Black Swans. Another aspect brought out in the movie is the possibility that Black Swan events may have a dual character, oscillating between black and white — indeed like the swan in *Swan Lake*. I am searching for the good side, but the bad side many be lurking there too. A striking example comes from the scientists who developed the atomic bomb (Oppenheimer, Fermi, *et al.*) during World War II. In their enthusiasm to save America and the Allies, they thought they were developing a good Black Swan. But the good had a dark side that these scientists (and society) slowly began to comprehend. Indeed, in this case, the Black Swan did truly have a dual personality. The nuclear weapon aspect hangs on as a global worry, while peaceful uses of nuclear energy, ranging from nuclear power to medical isotopes, have benefited and will continue to benefit mankind. What kind of Black Swan will you or I find (if we do) — a good one, a bad one, or one with a dual personality?

## Living at the Center of the Energy Crisis

After proposing that I write this book, Tom Valone later suggested that I consider updating the title of this book to something like "Life in the Center of the Energy Crisis." At first I hesitated to do that. But as I thought about it, I realized that I have been in the center of this crisis as I acknowledged years ago in my listing in *Who's Who in America*. There I said that unless we resolved the growing energy crisis, wars over dwindling resources will surely break out. As an individual researcher I saw the immediate problem, e.g., high gas prices, but I chose topics to work on that often were aimed at long-term improvements to energy sources. This is evident from this book's chapter titles, which range from aspects of fusion and fission reactors to nuclear batteries, fuel cells, and low energy nuclear reaction power cells. The research I have done and continue to do on these topics, however, concentrate on specific aspects of the technology that I could tackle as an individual researcher. Large companies with extensive resources must change their stance on research to overcome the energy crisis. However, their capability to change must ultimately come from new research developments obtained from numerous individual researchers like myself. Indeed, part of the energy crisis is bound up with the pressure stockholders put on companies to make near-term profits. This has prevented long-term research in many energy companies. Thus the attack on the "crisis" often seems more like a collection of temporary patches rather than a true fix.

The situation is further clouded because new technology such as renewable energy (solar, wind, biomass, etc.) is typically more expensive when first introduced than existing fossil energy, making it necessary for government subsidies to aid their entrance into the market. This entanglement of government and free market leads to confusion, because decisions about which technology to favor are very difficult. The United States government does not have an energy policy, only principles such as "energy freedom." Superimposed on that is the question of how to factor environmental issues into the market economy, e.g., the debate about carbon credits. Thus, the individual researcher can easily find all of this frustrating because the energy crisis can seem like a nebulous sea of confusion and debate. At the 2013 ARPA-E Energy Innovation Summit, the significance of the vast amounts of natural gas available in the United States as a result of fracking and horizontal drilling was impressed on me. Speaker after speaker, ranging from the Secretary of the U.S. Department of Energy Steven Chu to Blythe Masters, Head of Global Commodities at J.P. Morgan, emphasized the future impact this low-cost energy will have on the U.S. economy. My fear is that this concentration on natural gas may take emphasis from developing new energy sources; similar shifts happened before when oil prices suddenly dropped. When that happened, we became complacent and abandoned efforts to develop new energy sources. This time, supporters of energy innovation hope that cheap natural gas will not be viewed as a reason to discontinue addressing energy innovation. Regardless, natural gas will have a major impact on the situation. My hope is that despite projections that the natural gas supply can last for 100 or more years, we will recognize that this is still a limited resource when viewed over the scope of the industrial history of mankind. My personal approach, as documented in this book, has been to identify key research areas where I thought I could contribute, and then I put "blinders" on in order to enthusiastically seek that Black Swan without getting lost in the sea of confusion.

## Timeline and Apology

Over the years my search for a Black Swan has followed a zigzag path, whereby I have undertaken a variety of seemingly different research topics. These events are discussed in somewhat chronological order in the following chapters. Obviously some overlap time-wise makes it somewhat confusing. As an aid to the reader, a graphic timeline with footnotes is included in the Appendix. I hope this Appendix will add perspective for the reader about how these events fit together and the sequence of the research. When writing about past events, one's memory can easily be faulty. In my case I tend to remember and tell about the "good things." Certainly there have also been down times. And since I use names in association with events in this book, I know I have overlooked some events and people. There is just not sufficient space for everything. I offer my apologies for this matter in the Afterword.

# Chapter 1

# Early Days and Searching for a Starting Path

*Left: A recent photo of my office. Right: In my office in the 1970s. Note, how neat things were arranged back then! Forty more years of materials have come in since!*

I currently serve as an Emeritus Professor of Nuclear, Plasma, and Radiological Engineering at the University of Illinois at Urbana-Champaign. After 50 years at Illinois, I became Emeritus in August 2010. Now 79 years old, I could have retired some time ago, but I greatly enjoy teaching and research and wanted to keep at it. As Emeritus I still come in to my office every afternoon. I maintain my research labs, supervise students, and occasionally co-teach a course. Maybe, as Douglas MacArthur's farewell address at West Point suggests, old professors slowly fade away! I have tried to achieve a career characterized not only by innovation, but also by a willingness to consider new ideas and to employ patience and optimism in all endeavors. Many people and events helped prepare me for my search for a Black Swan event in energy development. In this chapter, I briefly recall the development of my interest in engineering through my childhood experiences and my parents' influence. It is impossible without writing another book to convey how much I owe my mother and father for steering me towards my career. Not that they ever tried to directly influence what I should go into, but they valued education and showed by their actions a deep faith, optimism about life, and a strong work ethic. I always felt boosted by my parents' support and confidence in me, even during the times when I seemed to screw everything up.

My father, G. Hunter Miley, received a MS degree in Chemistry from VPI, but his first job was a chemical plant construction engineer. He and my mother, Norma, married when he was in charge of construction of a chemical plant in Shreveport, Louisiana (where I was born). My mother was in her first year of teaching grade school, and they met because she happened to rent a room in the same boarding house where my father rented. Shortly after that, my dad was assigned an oil refinery construction job in Petrolia, Pennsylvania (about 50 miles North West of Pittsburgh). I was raised there. He was offered a job in the plant after it was up and running. Those were times of the Great Depression, but the oil refinery business remained robust, so my dad immediately took the job. Still, I think he longed to get back into the construction business after the economy recovered, but he never did. Plant maintenance, process updates, and expansion involved construction-like activities that he grew to enjoy also. Plus this gave him an opportunity to perform research for new processes for the refinery business.

*Some memories of my Dad (G. Hunter Miley) with rifle in hand at VPI (Left), and in front of the Chemistry Building at VPI, where he majored in Chemistry (Middle). The picture of me walking down Main Street in Shreveport, LA, with Dad during a visit with my grandmother remains a vivid memory in my mind (Right). I had resisted wearing the outfit I had on, but I loved walking with Dad.*

*To the left is a photo of a "Miley Retort," one of Dad's major inventions, near Hamburg, Germany. He traveled to a number of places around the world on plant construction projects. The stays were usually relatively short so my mother and I did not go along. One place he talked about as wild and beautiful was Aruba (Middle). I had long remembered this photograph in his scrapbook showing Dad (standing on right) with friends under a characteristic windblown beach tree in Aruba. In December of 2010 when Liz and I went for the first time to Aruba, I sought out a tree similar to the one my Dad stood by during*

*his visit (Right). I also sought out the oil refinery he worked on there and found it had been temporarily shut down in 2000 but was restarted later under new management.*

*My mother and father retired in Williamsburg, VA, and are shown here in 1968 when we visited with our son, Hunter. They enjoyed Williamsburg with all of its colonial activities and Dad's two sisters lived nearby in Yorktown, adding to the enjoyment. I used to visit my aunts in the summers when I was small — and fought the stinging nettles when trying to swim in the York River. There was a U.S. saltwater biological station there which I always visited. Inspired, I dreamed about becoming a biologist, but didn't. There is more in this book about my dad's technical work and his influence on me, but I cannot say enough about how my mother also influenced my life. She was always supportive and encouraged me when I had trouble and felt dejected, and she was determined that I get the best education possible. What luck to have had these parents!*

My father had a small chemistry lab in our basement where he studied his new ideas for chemical processes when he had time after work. I vividly remember my mother complaining about the chemical odors that sometimes came out of the lab. Our house was in view of the oil refinery where my father worked, and a chemical plant was also located near town. These two plants put out odors and soot from numerous stacks, forcing my mother to try to prevent me from opening the windows in my bedroom or elsewhere in the house. Jobs were the objective in those days, so clean air and the environment took a backseat. When lying in bed at night, we could hear oil wells creaking in the fields around the woods outside of town.

One of my dad's major inventions that came out of the home lab was a process for treating heavy sludge (created during the refining of crude oils) to form a coke for use in power plants. This "Miley Retort" was patented and sold by the L. Sonnborn Sons Oil Refinery of Petrolia, PA, to a number of other refineries around the world. I remember, for example, my father traveling to Germany and later to Russia to consult on retort start-ups. This creative atmosphere grew on me, and there was never any question in my mind that I wanted to become an engineer. Later I discovered physics and combined that as a minor subject area when majoring in Chemical Engineering at Carnegie Tech (now Carnegie Mellon University).

Felix Adler, then Professor of Physics at Carnegie Tech and later my colleague at the University of Illinois, taught my physics class and greatly influenced me. He was internationally known for his theoretical work on fast neutron nuclear reactors, designed to "breed" new fissionable fuel. His lectures about basic physics were always clear and inspiring. While he also attempted (and often failed) to demonstrate principles via experiments in class, his insights and abilities made a deep impression on me. Sadly, Felix died about 15 years ago shortly after retiring from the University of Illinois. The week before he died, Felix and I had an excited discussion about his plan for introducing a new physics course on fast reactors as an Emeritus Professor.

I signed up to attend Carnegie Tech during my junior year in high school at Mercersburg Academy when the Director of Admissions from Carnegie Tech came by on a recruiting trip. My mother had been insistent that I spend my last two high school years in Mercersburg Academy, which had a high reputation as a prep school. I had enjoyed my friends, basketball, and classes at Karnes City High School, the local consolidated high school near Petrolia. But my mother thought that a prep school would prepare me better to enter college than a small high school like Karnes City could. I left for Mercersburg somewhat reluctantly and was initially homesick. However things settled down quickly as I got involved in classes and other activities. I received a great experience and education there which did prepare me well for college, both academically and maturity-wise in the sense of living away from home and having to prioritize my activities and study time.

I must add that Karnes City High School, despite its small size, would no doubt have done a good job preparing me also. It had very dedicated teachers and the school kept course offerings to the basics. In fact, I recently had an interesting experience that relates to this issue. During a fusion meeting at Argonne National Lab some of the attendees and I were talking over coffee and I mentioned growing up in Western Pennsylvania. One person, Ray Sedwick, a young Assistant Professor from MIT (now at the University of Maryland) pressed me to say what the town I lived in and what high school I attended. When I responded Petrolia and Karnes City High, he said that is where he went to school and his father had been the principal there. Ray said it was and is a great school and felt it prepared him well for college. In any case, Ray is certainly an example of a successful graduate from Karnes City High.

Mercersburg has had its share of highly successful graduates also. One I should mention is the famous classic film actor, Jimmy Stewart. He too had been raised in a small town, Indiana, PA. Though I don't know for certain, I bet his mother, like mine, instigated his going to Mercersburg. He preceded me there by about 25 years and had already become a well-known movie star. It turns out that the desk he had used in one of the classrooms had been identified and was prized by many students for seating (desks were assigned in some courses). I vied with several others for it, but lost, ending up sitting next to the Jimmy Stewart desk. Some say that my voice reminds them of Jimmy Stewart's — sort of a middle Pennsylvania twang. But I still had a little southern twang superimposed and now have added a Midwestern accent.

Many teachers strongly influenced and helped me during my years at Carnegie Tech and throughout graduate school at the University of Michigan. Again, to describe all of these experiences would take another book, so I only mention a few events here. I have already mentioned the influence Felix Adler had on me. Another teacher, Herb Toor, later Dean of Engineering at Carnegie Tech, came to the school as a young Assistant Professor the same year I entered as a student. I was in his first class. I thought I was learning a lot, but was taken aback, along with my classmates, when the highest grade on the first exam was about 30 out of 100 percent! By the end of the course, the highest grade in the final was about 95, as I recall. (I received an A in the course despite it all.) New professors often expect more from classes than is realistic. I can proudly report that I learned from my experience with Herb — the highest grade in the first class I taught later at Illinois was 5 points higher than his had been!

I earned money in the summers when I was at Carnegie Tech to help cover my expenses. One year I worked at a local steel mill, the next year at a chemical plant, and the next at the Gulf Oil Research Laboratory. Interestingly, at Gulf Oil I was assigned to assist in the department studying advanced catalysis. However, after that, the focus of my research was such that I did not think further about catalysts. Now after 50 years I've been thrust back into the area as part of my recent fuel cell research effort. The understanding of catalysts in the early days when I worked at Gulf Oil was quite empirical. The situation has drastically changed now with improved diagnostics and advanced computational methods. Thus, recently when we needed to develop a specialized catalyst for our all-liquid fuel cell, we were able to make amazingly rapid progress by using some theory plus building on recent developments in the field.

Carl Monrad, Head of the Chemical Engineering Department at Carnegie Tech, took a personal interest in students in the department and had an open-door policy for them. When I was a senior trying to decide what to do and where to go for graduate school, I made an appointment with him. Monrad pointed out the exciting opportunities in nuclear engineering, a new and exponentially growing field at that time. He ended one of his discussions with me by saying, "Go West, young man." By "West" he was referring to his alma mater, the University of Michigan. Encouraged by that advice, I looked into Michigan and ended up there. As an NSF Fellow, I had a fairly wide range of possibilities, but Michigan had a renowned Chemical Engineering Department and had just begun a Nuclear Engineering (NE) Program (later to become a Department of Nuclear Engineering). Students in the NE Program were to be enrolled simultaneously in another department, and mine was Chemical Engineering. Also, Michigan had an excellent Physics Department. My minor in Physics at Carnegie Mellon had strongly influenced me. I seriously considered majoring in physics at Michigan, but I didn't. I did take a number of notable physics courses, including the Theoretical Physics series taught by Otto Laporte, a highly respected theorist. The theory in the course was challenging enough, but Laporte also kept us off balance with his idiosyncrasies. He always had an assistant stand by the blackboard to erase it with a full damp cloth after he filled it with equations. One time, he announced he would collect our notebooks and grade them based on that day's lectures. (I had good handwriting then, and always took detailed notes, so I received an A++!) Then, before the midterm exam, we learned he had burned his handwritten draft exam notes in a bonfire in the hallway, saying that it was a security measure he learned while he was a scientific advisor to the American Ambassador in Tokyo in the 1950s. I never learned if the fire alarm sounded, but I did see some dark smudges on the floor outside his door.

I never had a class from Don Katz, then Head of Chemical Engineering at the University of Michigan. "Dad" Katz, as he was fondly known by grad students, was an outstanding Professor who was simultaneously involved in community affairs (e.g., he was the head of the school board in east Ann Arbor). I had a number of classes from Stu Churchill, widely known for his work in heat transfer and thermal hydraulics. Stu was a great teacher and has a photographic memory for names. While now about 90 years old and Professor Emeritus at the University of Pennsylvania, he still attends most Michigan Alumni events. He frequently gives short talks naming attendees (who cover a range of years), often commenting in some detail on their research topics while at Michigan. He always recognizes me immediately and can bring up something personal about me that even I have forgotten. What a talent! I was not blessed with that great ability to recall names, but I do remember faces — names come with a time delay!

I was fortunate to be Joe Martin's thesis student. He was strongly interested in radiation effects on materials and chemical reactions, and that became my thesis topic. Joe was in Chemical Engineering, but was Director of the Nuclear Fission Product Research Laboratory and later became Director of the Technology Institute at the University of Michigan. I admired and have tried to copy, to some extent, his teaching style — a deliberate step-by-step development starting from the basics. Joe was also an avid tennis player. There was a rumor among the grad students that to graduate under him you needed to invent a machine that would renew the fuzz on an old tennis ball. As friends know, I also greatly enjoy tennis, but for some reason I never had the nerve to challenge Joe to a game. In retrospect, I regret that. I had the privilege of getting to know him and his family through various social events held at his house and local restaurants. I also admired his ability to be so diversified. Joe was an affiliate Professor in the Nuclear Engineering Program and also intimately involved in management of the new Ford Nuclear Research Reactor facility funded by DOE and by the nearby Ford Motor Company. This facility also housed extensive radiation-controlled laboratories ranging from a facility for biological and plant radiation studies to a hot cave for handling highly radioactive samples with robotic "hands."

I undertook an experimental thesis (with supporting theory) of radiation effects on select chemical hydrogenation reactions. This required building an elaborate high temperature chemical reaction chamber which was inserted into the beam port of the Ford Nuclear Reactor. I had an NSF graduate fellowship, so my support was covered. And, fortunately, Joe Martin had contract funds that covered the cost of the rather expensive components in this experimental setup. After many days and nights in the lab, the apparatus was working and ready to insert into the nuclear reactor beam port. But the start-up of the new Ford reactor faced several delays due to construction and licensing issues. I began to worry about when (and in darker moments "if") the reactor would ever start. Fortunately, my concerns were unfounded. It was commissioned only a few months behind schedule. My experiment then proceeded well except for a minor annoyance — a water leak along the beam port seal (fortunately not radioactive water) forced me to keep a mop and bucket handy! Over the years, I have met several other University of Michigan graduates who did experiments there and encountered this water leak which reappeared on and off.

I met my future wife, Liz Burroughs, at a graduate group in the Presbyterian Church on Wabash Avenue. She was a dietician at the University Women's Hospital. We were assigned to prepare the meal for the Sunday night dinner of the group, and I called her having planned the food (ingredients and amounts) to buy. As I had hoped, she was impressed! (Actually, she learned later that I consulted the wife of a friend for advice on this.) Liz's father, Robert Burroughs, was then Vice President for Research at the University of Michigan and I spent a number of great evenings at their house. My final thesis defense was in early November, 1958, and Liz and I married November 22nd. (As I commented earlier in the introduction to this book, I don't know what I would have done had I not passed the exam.) After a brief honeymoon trip, we returned and prepared to move to Schenectady, NY, where I took a job at General Electric's Knolls Atomic Power Laboratory (KAPL) in mid-December.

Our kids remember many visits and Christmas celebrations back in Ann Arbor over the years until Liz's father passed away. Then, her mother moved to Rochester, NY, where Liz's sister lives. Before going to Michigan, Liz's father, an experimental physicist by training, had worked for Kodak in Rochester, NY. He left Kodak to serve in the Navy doing radar development during World War II. When he left the Navy, he took a job as a manager for a department of General Electric Company (GE) in Schenectady. In those days, GE had the policy of moving managers periodically from facility to facility. Thus, Liz moved with her parents and sister, Lodi, to Schenectady twice, to Lynn, MA, twice (Marble Head and Nahant), to Philadelphia, and, most interestingly to me, Richland, WA. Her father was sent to Richland for several years as manger for the GE nuclear reactor project at the Hanford site under government contract. The reactors built there had a major role in the production of nuclear materials for World War II. Thus, Liz attended Richland High School, cheering for the famous "Bombers" sports teams. She remembers helping her mother stuff towels around door sills to prevent the blowing desert sand from seeping through the cracks in the "temporary" government home they occupied (which is still there). Liz went to Cornell University and then completed a therapeutic dietician internship at Duke University Hospital before taking a job at the University of Michigan Medical Center, where we met.

Like mine, Liz's parents nurtured and encouraged Liz and her sister. Also like my father, Liz's dad had an innovative/inventive nature with an electronics and woodworking shop in a spare bedroom. He loved to make and repair electronic gadgets and wood furniture. Bob had assembled his high-quality phonograph, amplifier, tuner and speakers from components, and he had a big easy chair placed just right for enjoyable "front row" full-volume listening. One electronic gadget was a non-lethal, electric contact that squirrels would set off if they tried to steal from the bird feeder. Squirrels would occasionally be seen somersaulting to the ground, temporarily stunned. When concerns about several house robberies in the vicinity came up, Liz's father constructed an elaborate sensor system to protect the house. It set off a horn located on a tree that would alert neighbors to a problem. Unfortunately, two flaws developed — an animal could set it off, and Liz's mom had trouble getting the hang of disabling it.

Liz's parents also loved to travel in their mobile home. Her father had equipped it with special electronics for various functions, such as sensing low tire pressure, an open refrigerator door, or sounding an alarm if too near to an object when backing up. When visiting us, her father enthusiastically looked for things he could work on and repair. And when he passed away some years ago, the people who bought their mobile home appreciated his many innovations. As electronics have advanced over the years, many similar devices are now offered commercially for cars and mobile homes. I learned much from Liz's dad and I was inspired by his creative bent. I tinker some around the home myself, but I remain a novice in comparison.

While Liz's father was in the Navy during World War II in Washington, DC, he brought a gaff-rigged Friendship Sloop. The family frequently sailed it on Chesapeake Bay on weekends. When the war ended and he accepted a job with GE in Schenectady they planned to sail their boat to a dock near the entrance to the Hudson River, an easy drive from Schenectady. On the way, as they sailed into the harbor at Atlantic City to spend the night, the boardwalk lights made recognition of buoy lights difficult and they ran aground on a ledge. The Coast Guard rescued them, and they spent the night in the Coast Guard Station. The tide came in the next day allowing the Coast Guard to drag the boat off of the ledge. They had to continue to Schenectady by rental van, temporarily leaving the boat in a shipyard for repairs. Despite adventures like this, Liz still loves sailing. Thus I brought a small boat that can hold two people. We sail off and on at a few inland lakes in central Illinois. It's fun, but nothing like having a Friendship Sloop on the Chesapeake!

## Further Reading

G. H. Miley and J. J. Martin, "The High Temperature Irradiation of the N-Heptane-Hydrogen System," *AIChE Journal,* Vol. 7, p. 593 (1961).

"Thermal Decomposition of Sludges," United States Patent 2897054, Hunter Miley, inventor, 07/28/1959.

# Chapter 2

## Burnable Poison Control for Nuclear Submarine Reactors

*The USS Seawolf used a pressurized water reactor while KAPL focused on boiling water designs. Indeed their first design used a liquid metal coolant-moderator to obtain a "fast neutron spectrum" type reactor. But Admiral Rickover finally stopped that line of designs, famously saying, "If the ocean was composed of liquid metal, GE would design a water cooled reactor!" I arrived at KAPL a few years after that decision.*

*Admiral Hyman G. Rickover is known as the "father" of the U.S. Nuclear Navy. He indirectly had an important influence on my career at KAPL and some lessons learned from his actions remain with me today.*

*(Photos from: http://www.history.navy.mil/bios/rickover.htm)*

My professional career began in late 1958 when, as a new PhD, I joined the staff of the advanced reactor group at Knolls Atomic Power Laboratory (KAPL) in Schenectady, NY. KAPL, run by GE, was one of the two research and development labs created in the United States to develop nuclear reactors for submarines (Bettis, run by Westinghouse in Pittsburgh, was the other lab). When I first walked into my office, I was thrilled to find that, among other things, my office colleagues were authors of the famous "Chart of the Nuclides" used worldwide by nuclear scientists. I was assigned the task of studying nuclear core lifetimes and finding ways to elongate their lifetime. This is obviously a key issue for long underwater missions, and I felt fortunate but intimidated by this challenging assignment.

Shortly after I arrived at KAPL, Admiral Rickover, the storied "father" of our nuclear submarine program, sent his right hand physics advisor, Alvin Radkowski, to KAPL to meet me. (The Admiral was renowned for the way he made contact with people working for him at all levels, including personal interviews with personnel hoping to enter the Nuclear Navy Program.) Radkowski had already earned a reputation as the physics "designer" of the Shippingport reactor near Pittsburgh, PA. It was designed by the Naval Reactors Group as a "first" demonstration of a commercial-scale, land-based power plant. As already noted, I was tasked at KAPL with nuclear reactor core burn-up studies (i.e., lifetime calculations). The hope was that such burn-up calculations, although severely stressing the then available computer capabilities, would predict ways to lengthen operation before refueling. As it stood, submarines were typically brought in early for refueling in order to be on the safe side. My calculations were done by sending punch card data to the David Taylor Model Basic facility in Washington, DC, to utilize their extensive computer facilities — among the best in the world at that time. Alvin Radkowski told me that Admiral Rickover feared I was a "young whippersnapper" who did not take core lifetime projections seriously enough! Consequently, the Admiral wanted me to spend a month on a land-based submarine training facility in West Milton, NY, where I would find out what would happen if the fuel burned up during a mission. Despite vigorous protests, I ended up at West Milton a week later. You did not say *no* to the Admiral! (Note: This is a "mild" Rickover story — many abound about this colorful "father" of the nuclear submarine development.)

When I returned from West Milton, I was asked to split my time on burn-up calculations with study of a new concept termed "burnable poison." The basic concept was to use a neutron absorber in the fuel rods such that it was transmuted by neutron induced reactions to a non-absorbing material over time, thus counterbalancing the burn-up of fissionable material. This was proposed to reduce the movement of control rods during burn-up. In view of the cramped space in a nuclear submarine, little space was available for control rods to come out of the core. Use of a burnable poison would counter this space problem. Researchers working on several subcritical facilities at KAPL wanted to study this concept but needed computation/theoretical support. I was tasked with selecting an appropriate element for the burnable poison and determine how to distribute it in fuel elements to provide the desired balance with loss of fissionable material. In addition to performing an intense series of calculations to study the problem, I closely collaborated with the experimental researchers to determine the best material for the burnable poison. That involved a trade-off between the poison's neutronic properties and metallurgical properties. After intense study, we came up with a solution to the problem that was eventually incorporated in the nuclear fleet. This breakthrough was reported in a series of still classified documents, including a series of theoretical computations I authored. The burnable poison technique has since been widely employed in both military and nuclear power reactors. While this discovery came close to being a Black Swan in the area of reactor physics, it pales into an immature Mute Swan when compared to great achievements by pioneers in the field, such as the landmark Fermi criticality experiment at Stagg Field in Chicago in the 1940s.

My rapid success in burnable poison development was possible because other great physicists at KAPL were already working on this issue. I just came along at the right time to pull together and synthesize the data. Still, this was a group effort and a group success. For me, the most important aspect was that this success gave me excitement and confidence — a real boost for a beginner. I am not sure what might have happened if the project had failed. I like to think I would have exhibited the characteristics of patience and optimism that I so value. Regardless, I retained my "free spirit" throughout and still longed to get into a research position where I had more freedom to work on fusion as well as fission.

My stay at KAPL was briefly interrupted when in the spring of 1960 I was called to active duty in the Army. I had been on the ROTC at Carnegie Tech and was commissioned as a Second Lt. when I graduated. This carried with it a two year active duty obligation. But I had been deferred for graduate school, during which time I attended Army reserve meetings. Thus I was promoted to First Lt. when I entered active duty at Fort Belvoir, VA. Due to my nuclear engineering experience, after completing basic training, I was assigned to the Army mobile nuclear power reactor group. A prototype reactor was in a test facility on base. It was designed for a small skid-mounted 100-kW reactor that could be airlifted into remote sites or to a base in the field. My active duty was terminated when the Army decided to release my whole class of new officers early to save money. The mobile reactor plant was eventually cancelled as nuclear reactors fell out of favor. Now, after many years, considerable interest has been expressed by several commercial companies to develop similar "Small Modular Reactors" (SMRs) for distributed power units.

My short time at Belvoir was a good experience, but after only six months I returned to KAPL to pick up the work on burnable poisons where I had left off. Indeed, finding a Mute Swan (burnable poison development) so quickly had only been possible because I choose to work at KAPL in the first place. In fact I had gone to KAPL after some indecision about where to start my career. I had also received job offers from several other laboratories. One was working on the DC-X fusion experiment at Oak Ridge National Laboratory (ORNL). I had already developed a keen interest in fusion, growing from the courses I took at the University of Michigan, so this was a very tempting opportunity to go into fusion research. However, several factors eventually led me to GE and KAPL instead. One of my criteria for selecting a job was to find a supervisor from whom I could learn. The position at GE was directly under Dr. Henry Hurwitz, who led a fusion effort at the main GE Research lab and was also a consultant to KAPL. He was known throughout the nuclear community for his brilliance (e.g., he was named by *Fortune Magazine* in 1954 as one of the top 10 scientists in the U.S. industry) and he was certainly someone who would be a great mentor. When I accepted the job, I thought I would have the best of all worlds, working with him and being involved in fusion and, to some extent, fission. GE had demonstrated a z-pinch fusion confinement experiment at the first unclassified International Atomic Energy Agency (IAEA) meeting on fusion 5 years earlier. Thus they seemed to be on the ground floor of the field. However, it took me a few months after accepting the position to complete my thesis and leave for Schenectady. In the intervening time, Hurwitz reported to the GE management that, in his opinion, practical fusion was too far in the distant future to satisfy their requirement for reasonably near-term financial returns on the research. Consequently, the fusion program at GE was cancelled. Thus, when I arrived in Schenectady, based on advice by Hurwitz, I was reassigned from the GE Research lab to KAPL. There I was placed in the advanced naval reactor physics group under Tom Frost, a senior nuclear physicist at KAPL. Frost's ability, insights and willingness to tutor a new employee, along with the great scientists in the group such as Gene Wachpress, Frank Feiner, Dick Dahlberg, and Norman Francis, made me quickly forget about my disappointment and I enthusiastically dove into the work. Dick Ehrlich, Head of the Advanced Concepts Department, was also very friendly and helpful.

Interestingly, years later, I unexpectedly met Dick Ehrlich at the University of Illinois in 1990 when he accompanied his wife, Eleanor, to a class reunion. Eleanor had been one of the very first women who graduated from the University of Illinois with an advanced degree. Eleanor was born on a farm in Illinois in 1918 and was the only woman in her 1938 Physics class at the university. She and Dick met sometime later at a physics conference after he received his PhD from Harvard. I had not realized this connection to Illinois when I worked for him at KAPL.

In 1961, when I had the opportunity to join the faculty at the University of Illinois, I decided to do it. My initial burnable poison work set the tone for my style of tackling new problems, with the goal of innovative solutions. My stay at KAPL had been brief, but it was a very important learning experience. In addition to the nuclear reactor technology itself, I learned much from my interactions with the great staff at KAPL. Also I learned discipline from having to report in detail all work back to Admiral Rickover's physics group in Washington. I gained new experience from working with the forefront computer systems at KAPL and at the David Taylor Model Basin laboratory in Washington, DC.

The time at KAPL, combined with my prior academic experiences, were formative periods during which I developed my personal goals for research and style for approaching new research problems. This style is best verbalized in the lecture I gave when awarded the Edward Teller Medal for innovative research in Inertial Confinement Fusion in 1995 (see the reference at the end of this chapter). My goal was and remains, as the title of this talk implies, "patience and optimism." Indeed, one can apply these principles to life as well as to research. We can be optimistic that things will work out, provided we turn our backs on the nagging desire for "instant gratitude" and take a long-range perspective. Thus, patience and optimism go hand-in-hand to guide one's life and work.

A key goal for my career has been on development of sustainable energy sources. This was verbalized best in the 1980s when I responded to the editor's request for a life's goal statement for *Who's Who in America*. I wrote: "My professional goal has been to ensure that future generations have a plentiful supply of economical, readily available energy such as offered by fusion (and renewable energy sources). Not only should this ensure a continued improvement in the standard of living for persons in all nations, but it should help maintain peace, which is threatened by the struggle to obtain and control limited natural sources of energy." Many others have expressed such sentiments in recent years as the growing energy crisis became apparent. However, the root causes of this crisis existed much earlier, even before I wrote the *Who's Who* statement.

## Further Reading

*Unfortunately, the classified works listed have never been released, but are provided for historical interest. For a general description of the concept see http://en.wikipedia.org/wiki/Nuclear_poison.*


G. H. Miley, "The Burnable Poison Depletion Code, M.S.F.O.," KAPL-M-GHM-1, Knolls Atomic Power Laboratory, (classified) Apr. (1959).

G. H. Miley, "WN-A Model S3G-Core 1 Life Study," KAPL-M-GHM-2, Knolls Atomic Power Laboratory, (classified) Feb. (1960).

G. H. Miley, "Analysis of Some Cores Designed for Long Endurance," KAPL-M- GHM-3, Knolls Atomic Power Laboratory, (classified) Feb. (1960).

G. H. Miley, "1995 Edward Teller Medal Memorial Lecture," *Laser Interaction and Related Plasma Phenomena, AIP Conference Proceedings,* Vol. 369, pp. 1334–1351 (1996).

G. H. Miley, "Nuclear Power Plants," Section 5.2; "Nuclear Power for the Future," Section 5.3; "Nuclear Fusion," Section 5.4, *Standard Handbook for Electrical Engineers*, Fifteenth Edition, McGraw-Hill, Inc. (2007).

N. Polmar and T. B. Allen, *Rickover: Controversy and Genius — A Biography*, Simon and Schuster, New York (1982).

# Chapter 3

# Nuclear Pulse Propagation and Fission Reactor Kinetics

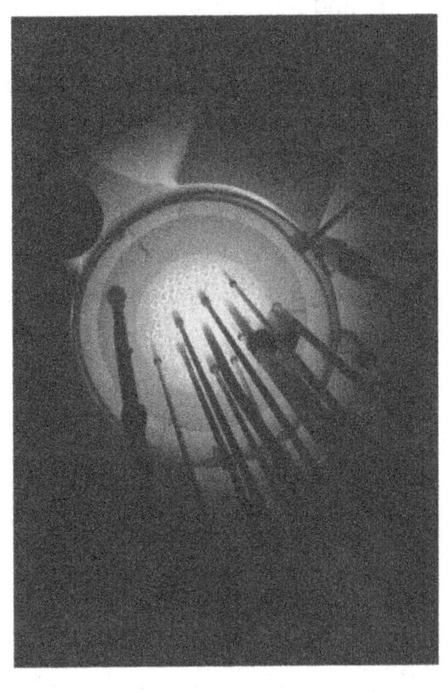

*Left: A view of the University of Illinois TRIGA reactor with a tour group on the stairs of the top deck of the reactor pool structure. Over the years, people on tours learned about nuclear energy, and a number of grade and high school students later remembered that exciting visit and came into our program. Right: A view looking at the core of the University of Illinois TRIGA Reactor as seen through the water tank. I spent many hours at this facility doing research on reactor dynamics, radioisotopes, nuclear-pumped lasers (NPLs) and radiation energy conversion. I was by far the main user of pulses in this TRIGA, and arguably in the United States. Anyone who was looking down on the reactor core like this during a reactor pulse saw an almost blinding flash of blue light (Cerenkov light created by high speed electron interactions with the pool water, the electrons being created from gamma ray interactions in the water). The visual effect was unforgettable. If I close my eyes, I can still see it!*

I left KAPL to join the faculty at the University of Illinois in 1961 as a fresh, new Assistant Professor to gain the freedom to work on fusion, and also because the new TRIGA research reactor (opened in 1960) offered so many research possibilities. This reactor was unique at the time, allowing safe pulsing of the reactor power from kW levels up to multi MWs in millisecond pulses. The neutron flux available for experiments also went up from ~$10^{14}$n/cm$^2$-sec steady state to ~$10^{18}$n/cm$^2$-sec at the peak of a pulse. This enabled many studies of reactor dynamics and various studies of transient radiation research in materials, chemical reactions, and biological systems. Pulsed operation was also the key factor that enabled the nuclear-pumped research I describe in Chapter 4. I wanted to participate in this forefront reactor kinetics research while also branching out into fusion and other new research frontiers. I thought the "freedom" offered in university research would provide that opportunity. It did in many ways, but still the demand to attain financial support for research in a field puts some unfortunate limitations on this freedom. I countered that problem by learning how to write good proposals to NSF, DOE, etc. Still, even to this day, despite my successes, I hate writing proposals. The success ratio for many agencies has dropped to less than 10%. It takes a thick skin to rebound from a rejection! One almost needs to do the research and then acquire support funding in order for a proposal to be successful. That is not my style. I want to propose pioneering research even if success seems risky.

*Left: During the 25th anniversary party for the Nuclear Engineering Program, as Chairperson, I presented the plaque to Ross Martin. As director of the Engineering Experiment Station in the College of Engineering at Illinois, Ross had provided the administrative leadership needed to establish the program. Right: At the party I also shook hands with Marv Wyman who, along with Felix Adler (not shown), was hired to start the Nuclear Engineering Program at the University of Illinois. Marv was instrumental in bringing the TRIGA reactor to Illinois and was the one who guided the academic program as well as handled hiring new staff. He was my mentor during my early years at Illinois. Later, Marv left to take a leadership position at the Jefferson Lab in VA, but shortly after that he passed away due to cancer.*

Thinking about the TRIGA reactor brings back many great memories. Over the years, I was the most frequent user of this facility. When Ron Martin, Head of the Engineering Experiment Station under Bill Everitt, Dean of Engineering, proposed formation of a Nuclear Engineering Program at the University of Illinois, they hired two key physicists to do it: Marvin Wyman and Felix Adler. Both had distinguished careers in nuclear and reactor physics. Marv was to create the experimental program. He had extensive experience on a research reactor at Los Alamos National Laboratory (LANL) and spearheaded the campaign to get the TRIGA reactor to the University of Illinois. Felix Adler had an international reputation for fast neutron fission theory. Marv hired Gerry Beck, a prior classmate of his during his own days as a graduate student in physics at the University of Illinois, to be in charge of the TRIGA facility. Gerry was a fixture there for many years and a great friend.

In addition to fast neutron reactor physics, Felix Adler was known for his work on nuclear cross-section resonance theory. He concentrated on creating several courses, such as reactor kinetics, and gathered a group of thesis students interested in reactor theory. When I arrived at the University of Illinois, I was initially somewhat awkward around Felix because, as noted earlier, I had known him as my physics professor at Carnegie Tech. I got an A in his course, but I was never sure he thought I deserved it. One memory of his class that I later joked with him about was that his in-class lab experiments never seemed to quite work.

Eventually Marv Wyman became Chairperson of the Nuclear Engineering Program. This was an interdisciplinary graduate program, so all faculty members were cross-listed with other departments. Only MS and PhD degrees were offered. Marv and Felix were naturally cross-listed with physics and went by that title. When I was hired in 1961, I was the first to have a Nuclear Engineering degree, so I was made Assistant Professor of Nuclear Engineering and Physics. Later, I was reassigned from Physics to Electrical Engineering, as I became heavily involved in plasma research in the Electrical Engineering labs and taught several of their courses.

One interesting quirk of events was that Jay Brownlee, a friend of mine and Liz's from the graduate group at the Ann Arbor First Presbyterian Church (where Liz and I met), graduated a little ahead of me and went to Illinois as Assistant Professor of Chemistry. He joined what was then a very active group working on nuclear chemistry. Thus later when I began looking for an academic position, Jay found out and contacted me. He enthusiastically encouraged me to come to Illinois. Jay was loosely connected with the Nuclear Engineering Program because he was involved in radiochemistry along with John Hummel. When I came to Illinois for my interview, Jay took time to meet separately with me and show me around the University and surrounding area. He was a strong influence on my decision. When we moved to Illinois, Liz and I were close friends with Jay and his wife, Janet. Jay remained at Illinois about ten years after I arrived, but then left for a government job in Washington. Part of the reason was due to a decreasing emphasis in Nuclear and Radiochemistry at Illinois. That was a trend at universities over the country then, but in recent times there has been somewhat of resurgence in the area. However, with the reactor gone from Illinois (used for many experiments by radiochemists like Jay) we have not participated in this trend.

Interestingly, a year after I came to Illinois, another friend from the graduate group in the First Presbyterian Church of Ann Arbor, Dave Eades, joined the biology faculty. Dave and his wife Jane had also met in the graduate group. He was involved in DNA studies and using advanced computational methods to extract data from studies of grasshoppers. I visited his lab several times and was very impressed. However, after a few years Dave changed directions completely, and left the University to become a housing developer. His wife was very experienced in business and the two of them successfully developed several major housing communities in Champaign. I am not sure why Dave decided to go into this business, but he was obviously good at it. Jane passed away some years back, but I still see Dave at our Champaign Presbyterian church.

The concept of an interdisciplinary program like ours was not well understood at the time, both within and outside of the University. It was the vision of Dean Everitt that the entire Engineering College structure might eventually follow our lead. He felt that the departmental structure created artificial "walls" between people. Recognizing the problem, some other universities such as UCLA had already converted from departments to a specialty structure. Dean Everitt was heavily involved in the American Society of Engineering Education (ASEE), which was founded at the University of Illinois. ASEE had many sessions on the subject. But the movement to abolish departments never gained traction. Today the University of Illinois is still strongly departmentalized with centers and institutes superimposed. In our case, the Nuclear Engineering Program was converted to a Department in the 1980s when the bachelor's degree was added to the existing graduate degrees. A strong motivation to do that came from the competition for faculty and students we faced from other universities who told people that a "program" was not as "important" or did not have the "standing" of a department like they had.

Aside from all of that, the mechanism of an interdisciplinary program at the graduate level was a very effective way to start the field at Illinois. It brought in great colleagues from other departments who wanted to apply their expertise to nuclear engineering. I learned much from them. Early people involved included: Bei Chao (and later Tad Addy) of Mechanical Engineering; Tom Hanratty of Chemical Engineering in thermal hydraulics; Ben Ewing and Chet Seiss of Civil Engineering in environment and earthquake structural engineering; Dan Hang of Electrical Engineering in engineering economics; Bob Bohl of Metallurgy in structures and radiation effects on materials; Art Boresi, Chuck Taylor, and George Costello of Theoretical and Applied Mechanics in material dynamics; John Hummel and Jay Browlee in Nuclear Chemistry; Bob Twardock in the radioisotopes area of Veterinary Medicine; and Howard Ducoff from Biophysics in radiation effects on biological systems. Marv Wyman and Felix Adler were core members, but both also held their appointments in Physics. Ross Martin, in the Dean's office as head of the Engineering Experiment Station continued to provide strong administrative support and guidance for this new "Program" of Nuclear Engineering. Among the affiliated staff, Bei Chao, Bob Bohl, and Dan Hang were particularly involved in the Nuclear Engineering Program, teaching core courses and attending staff meetings. Ben Ewing also frequently attended, and was very effective by calming things down when discussions started to become heated. Many affiliated staff attended summer workshops on Nuclear Engineering at Argonne National Laboratory (ANL) to gain more insight into this new field and understand better how their specialties fit in. All of this enthusiasm made the NE program vibrant and fun to be a part of.

These colleagues plus Marv, Felix, and the others nurtured my early years in academia (other faculty, not listed here, came along later to influence my career). I could recount many stories from those days, but one I remember well is that during my first year at the University of Illinois, an important conference on fission reactor dynamics came up at the University of Arizona in Tucson. I desperately wanted to attend, but didn't have a research contract to support the expenses. Marv Wyman, who was my mentor in those days, came up with a scheme to hire a University plane. (Many planes were used by Illinois in those days to transport faculty around the Midwest for extension courses. They were part of a fleet housed at the Institute of Aviation at Willard Airport. Willard was also used for commercial flights by Ozark Airlines, but is owned by the University.) Marv and Felix had money to support hiring the plane. His scheme was that I could join them and thus save airfare — a kind move to help a junior staff member. The plane came with an experienced pilot and was a small two engine craft. All went well, except that on the first day we landed for the night about 300 miles north of Tucson, just over a mountain range. Unfortunately, the next day, with some downdrafts over the mountains and our plane close to the weight limit, we had to divert to find a lower pass through the mountain range. Consequently, we arrived a half a day late, missing the important opening keynote sessions. Felix and Marv could have gotten there on time on a commercial flight, but I think they forgave me for this.

Some years later, Marv stepped down as Chairperson of the Program and recommended me as his successor. Felix retired while I was Chairperson, and came to my office with plans to still teach a course and write a book. I enthusiastically agreed and we shook hands as he left my office. The next week, he unexpectedly passed away. Art Chilton, a former Naval Officer who retired from the Navy to become Professor of Nuclear Engineering (also one of my best friends and mentor who is discussed further in Chapter 18), also passed away shortly after retiring with great plans. And, as I pointed out earlier, Marv Wyman also passed away shortly after leaving Illinois. He had planned to work for a few more years at the Jefferson lab. I hope that I can avoid continuing this "tradition" and, despite having recently retired, keep going strong for many years in my search for a Black Swan.

My initial research effort on the pulsed TRIGA reactor was focused on fundamental studies of neutron pulse propagation in various neutron moderating media such as graphite, heavy water, and light water. The point is that the neutrons produced in the reactor during a pulse have a high speed but still undergo a significant (millisecond or so depending on distance of travel) time delay as they diffuse through these various moderators. My first study considered propagation through a graphite block contained in the reactor that served as a "thermal column" to thermalize (slow down) neutrons and direct them onto test samples or test assemblies located at the face of the thermal column. While the neutron propagation process seems very academic, understanding it is a key to explaining the dynamics of large fission power reactors. In that configuration, various bundles of fuel elements across a core "communicate" with each other by such neutron diffusion. This process is fundamental for determination of the time scale for power changes during power excursions caused by rapid movement of the control rods or by a fast accident mechanism. In subsequent work, P. K. Doshi, my first PhD student, and I built a second fission reactor core which was coupled to the main TRIGA core through a graphite block. It used "used" fuel elements I obtained on "surplus" from the reactor facility at a Medical Center in Bethesda, MD. We then studied the dynamics of this coupled core configuration when the main TRIGA core was pulsed. Interestingly, the second core, which ran at only a few kilowatts, was named the LOPRA (Low Power Reactor Assembly) and was licensed through the Nuclear Regulatory Commission (NRC) as a separate reactor (LOPRA is further discussed in Chapter 18). I vividly recall the licensing inspection. The inspector spent a day studying the setup and finally called us in to say that he would recommend the license if we would move the control instrumentation (meter, power supplies, switches, etc.) from sitting on a table to a rack. He said that all reactors used panels (racks) for controls! We had been happy with the table set up, but rapidly agreed. All of this went quickly, over a few months, but I am sure such licensing would require an order of magnitude more time today as regulatory procedures have become more complicated with numerous added steps and requirements. Unfortunately, in the early 2000s the University of Illinois decided to shut down the TRIGA reactor because of unfounded concerns about financial liability due to potential legal suits, possibly initiated by students passing by the facility every day, or someone in a nearby classroom or office. Thus both reactors, the TRIGA and the LOPRA were officially decommissioned.

Very talented and dedicated students joined me in these early neutron efforts. They include P. K. Doshi, Hassan Hassan, Gary Thayer, Larry Miller, K. Y. Cheung, and Harold Kurstedt. All continued on to great careers themselves. My main contact with former students has come through seeing each other at American Nuclear Society (ANS) meetings. Both P. K. and Hassan gained high management positions at Westinghouse and Babcock & Wilcox, respectively. Indeed, in later years, various students seeking employment asked me if they could be introduced to P. K. and Hassan. P. K. has always been a social leader and served informally as a contact to get various graduates in his class together for alumni events (plus some weddings). This not only included graduates but eventually their families as well. Harold Kurstedt ended up as Professor and Head of the Nuclear System Institute at Virginia Polytechnic Institute (VPI) in Blacksburg, VA. Also, he has shown Liz and I photos of a beautiful old farm house in the mountains near Blacksburg that he restored for his home.

Larry Miller did a pioneering computational study involving development of a special Monte Carlo code using a perturbation method to study reactivity coefficients in Fast Fission Reactors. Then, some years later, he became the president of the modest-sized Savings and Loan Bank in a Chicago suburb. Indeed, it is not uncommon for a person with extensive math/computational background to migrate into various financial businesses. Larry was the first, but not the last, of my students to do that.

As I discuss further in Chapter 18, during this time period a pioneering experiment on the TRIGA was instigated by a question from Harold Kurstedt, a student in my reactor physics course. He asked, "How rapidly can we pulse the TRIGA reactor?" This resulted in the first exploratory rapid pulsing experiments which eventually led to a research program on this subject. Fortunately in those days very little paperwork was necessary to propose a new experiment on a research reactor. I was a licensed operator on the TRIGA so we did not need to take time to contract with someone to operate the reactor for this experiment. Thus, it was done within a week. Today, the paperwork required to approve the experiment would potentially take many months and, in fact, might never be approved.

The experiment produced results that were not anticipated in advance, but are obvious in retrospect. Namely, the fuel heating during the first pulse causes the second pulse to have a lower peak power due to the negative temperature coefficient. However, the lower peak power of the second pulse resulted in less net temperature rise so that the third pulse actually had a higher power. These variations continued, but soon an equilibrium pulse height was obtained. We had not anticipated that behavior, but in retrospect immediately understood why it occurred (a Black Swan discovery by definition). Such a strange pattern might appear frightening to a first time observer. Fortunately, there is not a true safety concern since the peak powers are always bounded and less than the initial. The built-in safety mechanism involved is due to the inherent negative temperature coefficient associated with use of special ZrH-Uranium fuel elements in the TRIGA reactor design. This is also the basis for achieving the remarkable pulsing capability of all TRIGA reactors.

## Further Reading

H. Hassan and G. H. Miley, "The Period Effect in Reactor Kinetics," *Trans. ANS*, Vol. 9, pp. 466–467 (1966).

H. A. Kurstedt, Jr. and G. H. Miley, "On the Roots and Coefficients for the General Solution of the Reactor Kinetics Equations," *Nucl. Sci. Eng.*, Vol. 38, pp. 80–82 (1969).

H. Kurstedt, Jr. and G. H. Miley, "Short-Interval Series Pulsing Experimental Studies and Numerical Experiments," *Nucl. Tech.,* Vol. 10, pp. 168–178 (1971).

H. Kurstedt, Jr. and G. H. Miley, "A Temperature Formulation of the Space-Averaged Reactor Kinetics Equations," *Nucl. Sci. Eng.*, Vol. 43, pp. 319–327 (1971).

G. H. Miley, "Reactor Pulse Propagation," *Nucl. Sci. Eng.,* Vol. 21, pp. 357–368 (1965).

# Chapter 4

# Nuclear Pumped Laser (NPL) Research

*My work on NPLs led to memorable collaborations and visits to Russia even prior to the collapse of the Soviet Union. In my first trip to Russia after the end of the Cold War, the Russian NPL scientists disclosed that they had held "All-Russian" NPL conferences each year, rotating among locations at Arzamas-16 and Chelyabinsk-70, the two main secret cities in the U.S.S.R. I was presented with this photograph of attendees at the first such meeting. The senior scientists who attended were some of the top laser and nuclear scientists in Russia. Pictured in the center of the bottom row is A.A. Sinyanskii (beard and moustache), and A.N. Sizov is to the right of the woman in white raincoat. S.P. Melnikov is in the dark glasses behind and above them. E.P. Magda is in the light coat situated to his left. S.I. Yakovlenko is in the upper row, fourth from the left of E.P. Magda. P.P. Dyachenko is situated at the far right in the upper row (white raincoat and dark tie). A.V. Zrodnikov is sixth from the right of Dyachenko in the upper row (his head and scarf are visible). A.M. Voinov is the farthest right in the first row (his hands are in the pockets of jacket, and he is wearing glasses).*

*Alex Filyukov, shown to the left with his prized hunting dog, enjoyed a "privileged" position with a large flat in Moscow, a private car, a dacha on the Black Sea, and freedom to go hunting with his dog in the nearby wooded hills. I first met Alex in the early 1980s at an International Plasma Physics meeting in Moscow. He was exceedingly friendly during discussions. I realized that he knew almost everything about my research that had been published, suggesting to me that he had been assigned by the KGB or someone to question me. Later, he invited me to his apartment, where this photograph was taken. As time went on, Liz and I became friends with Alex and his wife, Luba. After the fall of the Soviet Union, Alex gained more freedom and wrote in an email he was very happy. But his happiness soon disappeared as chaos developed in the new Russia. He no longer enjoyed the privileges granted him by the Communist Party. His living conditions rapidly declined and he had to give up his big car. When he went to his dacha one weekend, he discovered that looters had torn most of the wood siding off, probably for firewood. He died a few years later at a relatively young age, perhaps due to years of drinking vodka and smoking. Indeed, life expectancy for males in Russia remains very low for a developed nation.*

*Some years after I first met Alex, and after the Soviet Union fell, he made his first trip out of Russia to visit me. Prior to that, officials there had forbidden him from leaving their sight due to his weapons knowledge, including their secret work on NPLs. He included a visit to Illinois and as part of that we held a party for him in our house. Some ten years earlier during one of our visits to Moscow, Liz and I had admired the samovars on a shelf in his*

*house. They had been handed down in his family. Alex insisted that we take one home, and would not take "no" as an answer! He gave us the newest one, labeled 1928, because removing "antique" items (made before the Russian revolution of 1917) was forbidden by law. After bringing the samovar home, we were unsure how to properly light it, so we kept it unused on display in our dining room (and still do). Alex was delighted to demonstrate the proper technique by lighting it up to make tea for the party at our house. In the photo we see him bent over it on our patio with smoke starting to emerge. He stuffed some dry leaves and twigs in along with folded paper to get it going. For a draft he used a fireplace bellows to blow air through the samovar's grated back. As the water in it was heating, one of our guests said that there was a fire truck coming down the street — looking for the source of smoke that one of our neighbors had called in about! During this visit Alex said he wanted to leave the weapons area completely and apply his mathematical abilities to studies of human genome sequencing. Indeed, he had already come up with a new approach which he had been discussing via email with several experts in the area at the Illinois Institute of Technology in Chicago. He subsequently visited them and at the time of his death was a co-PI on a proposal they had jointly submitted to the U.S. Institute of Health.*

*Left: I was one of the first "foreigners" permitted to visit the NPL research facilities in the secret cities after the fall of the Soviet Union. Shown here is a NPL set up, along with detectors and other equipment in front of a special pulsed research reactor built for NPL research. Right: There were about five specially-built research reactors designed for NPL research in the secret cities. Here is another pulsed system with large laser tubes located on top of the structure standing on posts in the foreground. The room incorporated openings through a shielding wall with diagnostics behind the wall, reducing radiation interference with detectors.*

*Left: In Arzamas-16, I was shown the Russian nuclear museum. This museum recounted the history of the development of the nuclear bomb and the hydrogen bomb in Russia (much like our museum at Los Alamos National Laboratory, now the Bradbury Science Museum). Here, my tour guides are standing in front of a mock-up of an early nuclear warhead. Right: The museum contained an amazing number of nuclear warheads with plaques describing them. Each looked frighteningly "ready-to-go."*

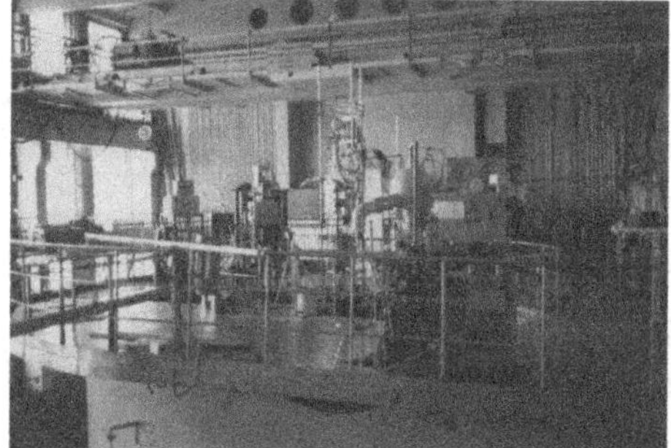

*Left: I made four trips to the U.S.S.R. prior to the fall of the Soviet Union. These were for discussions of NPLs, at the invitation of the Russians, hosted by Alex Filyukov and scientists from the Lebedev Physics Institute in Moscow. I knew that Alex Voinov had been doing experiments at a laboratory called MITI somewhere north of Moscow, but was not permitted to visit it. Later, after the end of the Soviet Union, I was invited to come again and was hosted by Alex (second from left), and observed his NPL facilities at MITI. He is shown holding one of his laser cells. Right: The top of the research reactor at MITI which Alex Voinov used for his early NPL experiments. It contained special neutron beam ports and diagnostic equipment for this research. Lasers developed here were then sent to one of the high flux facilities at the secret cities for further development.*

*Left: There were few cars in the secret cities (or even in Moscow) in those days. Still streets in the secret cities were very wide. I know the situation has changed considerably, including traffic jams in Moscow, but I haven't returned in recent years. Right: I visited Moscow State University several times. On the first, prior to the fall of the Soviet Union, I was to give a lecture, which as I describe in this chapter, was cancelled when they decided that the subject (NPLs) was "classified"!*

Before discussing my experiences in Russia, I should first explain how this nuclear-pumped laser (NPL) research started, and the events that occurred as it moved forward. In 1963 I attended a summer faculty institute on "fast" fission reactors held at the national reactor test site near Idaho Falls, ID. ("Fast" reactors use high energy, or "non-thermalized", neutrons to cause fission, as opposed to thermalized neutrons used in light water reactors. Light water systems are currently the dominant type of nuclear fission power plants worldwide. Fast reactors are favored for future "breeder" systems to generate more fissional fuel than consumed.) This summer institute involved travelling to the fast breeder reactor located in the desert (about 60 miles from Idaho Falls) on a government bus. During the long rides back and forth, I studied the first textbook that had been published on lasers. Indeed, lasers had only been discovered a few years earlier, though Einstein had written a fundamental paper about population inversion and stimulated emission some 46 years earlier. The theory developed by Einstein is fundamental to lasers, and should have sparked their development earlier, but scientists of the time failed to understand its significance. Thus, while one of Einstein's great achievements, it failed to have the immediate effect characteristic of a Black Swan event. Rather I guess it can be considered as a "time-delayed" Black Swan! The so-called "Einstein Coefficients" are involved in all theoretical analyses of lasers, and the effect lasers have had on all of our lives is truly immense.

After reading the laser textbook, the thought dawned on me that instead of using electrical energy to drive the laser, why not use nuclear energy? This would then be a nuclear-pumped laser (NPL). This thought had a profound effect on my research over the next decade. Upon returning to Illinois, I immediately began planning a NPL experiment.

This experiment was to use the pulsing capability of the TRIGA reactor in order to obtain the high neutron flux needed to deposit adequate energy into the laser medium. Neutron energy would be converted to kinetic energy of alpha particles via reactions in boron-coated walls on the laser tube. These alphas would in turn deposit their energy in the laser medium to drive the laser. Initially, I thought that I had discovered this process, but soon found that Lloyd Herwig of the United Aircraft Corporation in Hartford, CT, had written several internal company memos concerning a similar process. Further, Herwig had carried out some initial experiments (but had not yet achieved lasing). Based on Herwig's work and on my own study of the laser process, I selected carbon dioxide, $CO_2$, as the NPL medium. This review to choose the laser medium was hampered because the detailed laser kinetics was not well developed at that time, even for conventional electrically pumped media. My logic was that the energy deposition threshold for lasing decreases with wave length and $CO_2$, with its infrared output combined features of a low threshold and very high conversion efficiency (due to good energy mating between pump energy and the energy level of excited state) compared to other lasers. Thus, I estimated with crude calculations that a $CO_2$ laser should easily work with nuclear pumping, even under steady-state operation of the TRIGA reactor without resorting to pulsing to get high intensity neutron pulses. I liked the fact that this gave a big "safety margin" for errors in the calculation. If steady-state operation did not succeed we could always go to pulses to achieve a three-fold increase in neutron flux. I formed a close collaboration with Joe Verdeyen, a colleague in the Gaseous Electronics Laboratory at the University of Illinois, who had already built a novel electrically pumped transverse discharge $CO_2$ laser. This added great expertise to the project. However, our attempts to create lasing in $CO_2$, which lasted over a year, ended in failure. As a "consolation prize," Joe and I did manage to use nuclear pumping as a way to enhance the output from an electrically driven transverse discharge $CO_2$ laser. (This is typical of many times where I tried to turn a "failure" into a "success", i.e., a Mute Swan sighting.)

While working with Joe Verdeyen I got to know a number of staff and students working in the Gaseous Electronics Laboratory. The group greatly influenced my work and knowledge of gaseous electronics and lasers. The lab had been constructed as part of the University's effort to lure leading gaseous electronics scientists to join the EE faculty. One of these was Ladislas Goldstein, who was later appointed to be the director of the lab. Ladislas had a truly fascinating background. He began working at Marie Curie's laboratory (the Radium Institute) in 1927 and, when Marie died in 1934, he completed his PhD under her successor. I was fortunate enough to attend a fascinating seminar he once gave recounting his graduate student days in the Radium Institute. Marie Curie is a widely recognized example of a person giving almost complete dedication to her goal of a Black Swan.

Goldstein was fascinated with the physics involved in radiation effects on plasma discharges, but was not convinced that a NPL was possible. He had several students undertake experimental studies of radiation effects on discharges using the TRIGA reactor. These really probed some of the basis physics aspects of nuclear pumping. But their research goal was not an actual laser like I wanted to develop. Thus, during my visits to the Gaseous Electronics lab, Ladislas and I had a number of discussions about NPLs.

Ladislas always brought up new unresolved issues that might block lasing. Thus I came away challenged to think deeper about the subject. While that seemed frustrating during the discussion, I learned to appreciate the challenges he presented, and that was in the end a real help in my quest for achieving NPLs.

While not directly related to NPL issues, I recall vividly one "bad" Black Swan event in Goldstein's life. When he retired, Goldstein and his wife decided to return to France. They arranged to ship many personal belongings back by a freighter ship scheduled to depart from New Orleans harbor. Over the years, despite much urging by colleagues, he had never written a book on gaseous electronics. To the pleasure of colleagues, he announced that he was taking his collection of lecture notes to France to prepare this long anticipated book. Unfortunately, these notes and other papers were packed along with clothes for shipment on the boat. In a freak accident the freighter sank outside of New Orleans and his notes were lost. He had no copies of them. So the community lost a priceless store of knowledge that he had gained over many years of forefront research. His research papers remain, but this book would have pulled many loose ends together.

Another staff member in the Gaseous Electronics Lab whom I got to know well was Blake Cherrington. Blake collaborated closely with Joe Verdeyen and often co-taught the EE gaseous electronics and laser classes with him. Because I did not have a background in gaseous electronics I wanted to come up to speed quickly. I did not have time to attend these lectures, but I obtained the class notes. As I read them, if I had questions, I would go see Blake. He always graciously made time for me. I also had discussions with graduate students like T.V. George and Gary Eden, who were doing their thesis projects on gaseous electronics. (T.V. later joined the DOE in Germantown, MD, and served in the Office of Fusion Energy as a project manager. Gary went to the Naval Research Lab in Washington, DC, but later returned to Illinois and is currently Professor of Electrical and Computer Engineering.)

Over the years, others have also tried nuclear pumping of $CO_2$ without success. In retrospect it is clear that the difficulty encountered arises because nuclear radiation causes excessive disassociation of $CO_2$ bonds, inhibiting coherent vibrational action needed for lasing. This "mistake" of picking $CO_2$ caused us (colleagues, students, and me) to lose the honor of achieving the first NPL in the United States. That honor goes to David McArthur of the Sandia National Laboratories (SNL) who succeeded in driving a carbon monoxide (CO) laser using cryogenic cooling of the laser medium along with nuclear pumping. However, despite this initial frustration, my students, along with other collaborators, and I went on to achieve a number of other "firsts" in this field. (Again, a Mute Swan accomplishment.) These include the first "minority species" NPL and the first NPL with output in the visible range. In addition, we laid the groundwork for nuclear pumping of excimer lasers and for nuclear pumping of the iodine exchange laser.

When I originally started NPL research I was mainly enamored with the thought of directly turning nuclear energy into optical energy. Thus, my main focus was the physics and engineering of NPLs. Both the students working with me and I sensed that there were great scientific discoveries to be made, and the TRIGA reactor at Illinois provided a unique facility to do that. I had not thought much about possible applications beyond a vague vision of space power beaming. We had a strong desire to be first to discover new types of lasers using nuclear rather than electrical pumping. Like standard electrically-driven lasers, NPLs can use a wide variety of laser media, giving a wide range of power levels and output wavelengths. Thus there are a variety of types of NPLs possible, analogous to the many different electrically-pumped lasers, ranging from the small HeNe laser pointers to the massive Nd-glass laser for fusion experiments at the Livermore National Laboratory. We thought that applications would easily follow as knowledge grew, but frankly did not anticipate that a prominent one would involve the *Star Wars* program! In any case, we managed to make a number of new and important contributions to this field. These successes were enabled by the ability of the TRIGA reactor to deliver ultra high neutron fluxes during the 20–30 millisecond pulses. Also, the Illinois TRIGA had a unique beam port (i.e., a hollow tube where experiments could be inserted) that passed through the shielding and the pool water along the face of the reactor core. This beam port allowed alignment of the laser cavity using a laser directed through the port from end to end (i.e., one side of the reactor to the opposite). This was a distinct advantage because NPL alignment can be very difficult otherwise. Further, instrumentation could be brought in from one side of the port and the experimental laser beam observed from the other.

The TRIGA reactor was shut down for upgrading during our fourth year of NPL studies. During this time, Jean Guyot, one of my earlier PhD students, and I worked at the fast burst reactor at Oak Ridge National Lab (ORNL). Jean was doing his thesis on NPLs and we were "sure," according to my "best" calculations, that a HeNe laser would work, assuming impurities in the laser mixture could be kept exceedingly low. Preventing impurities was a demanding task, requiring a bakeable all-glass system. Jean was French and seemed to have inherited their knack for working with glass. He essentially taught himself (with some help from the glass blower in our chemistry department) to expertly blow glass. He constructed a masterful all-glass laser and gas filling system for the project which we transported to ORNL in the back of my station wagon. Things first became dicey when guards at the ORNL gate singled out Jean for extra security checks despite his having filed papers with their security office in advance. (Jean was a proud Frenchman, displaying a large picture of Charles de Gaulle, then President of France, over his experiment at the TRIGA.) He became convinced that they "had it in" for the French. We finally obtained a gate pass, but when we unpacked our equipment in the lab adjacent to the burst reactor the ORNL staff stopped Jean from torching the glass laser connection to the glass filling station (the two parts had been transported separately for safety). This, they emphasized, was a job for a union glass blower (ORNL support workers were generally unionized). When the ORNL glass blower finally showed up over an hour later, Jean rushed over to me excitedly. He said, "Don't let that elderly man near my glass system, he is too shaky!" I calmed Jean, telling him not to worry because the glass blower knew what he was doing, obviously having blown glass for many years. Just then, the glass blower slipped on a wet spot on the floor and broke the glass connection arm off the unit. Jean was horrified, but regained his composure and eventually repaired the break. Jean eventually forgave the glass blower, and me.

The next event played on our nerves even more. As it happens, the ORNL pulsed reactor was frequently used for health physics experiments. The experiment run ahead of us involved pulsed neutron irradiation of several donkeys chained to rails near the reactor (no doubt to interpret the effects of radiation on them as a surrogate for humans). We watched on the closed-circuit lab TV screen which read out from a camera focused on the reactor area. The donkeys kicked wildly after the burst of radiation from the reactor pulse. It was not clear to us if what the neutrons did was serious to their health, but I was left with a queasy feeling. Of course, the health physics experimentalists were accustomed to such events, and did not show any emotion — all business. Meanwhile, our experiment ran into problems. We could not obtain a reference laser to align our laser cavity (lasers were scarce at ORNL in those days). As a result, our alignment had to be done by eye, a less precise method. To avoid radiation effects on the detector, we had to locate the detector in a separate building some kilometers away and use reflecting mirrors to form a zigzag path that the laser beam, but not radiation, could follow. However, that setup made the detector alignment very tedious. We obtained several neutron pulses that day, but the detector did not record a signal. It was not clear if the experiment failed or if the detector was not correctly aligned. We would have kept trying but our assigned reactor time expired. More donkeys moved in to replace us. So, we packed up for the long journey home.

That experience had been frustrating, but Jean went on to get some great physics data on the radiation-induced HeNe plasma system using our TRIGA when it came back online. The added time Jean spent waiting for the TRIGA upgrade had one good benefit, giving Jean time to meet and marry a local girl, Kate, who was a high school French teacher. Later they went to France where, as I note in the chapter on ICF, Jean took a job with the French General Electric Laser group. He came back into my life when, as the French General Electric Laser representative, he helped set up a large Nd glass laser they sold to KMS fusion in Ann Arbor, MI. Subsequently Liz and I visited the Guyots at their house outside of Paris several times during trips to Europe. I need to go again, because Jean beat me at tennis the last time and I made him promise a rematch. Also their French-style oversized cups of coffee and cream (or better said: cream and coffee) are so good!

Another area, spearheaded by the thesis work of Mark Prelas, that we pioneered was the so-called "impurity" NPL (so named because the collision transfer of energy from the main pumped state to a minority species causes lasing in the latter). Because the input power went into the majority species, a large "pool" of excited states was available to transfer energy to the upper laser level. This feature allowed lasing with very low input pumping power, enabling a NPL with a record low input power threshold. Later, one of my students, Yasser Shaban, achieved a remarkable visible output NPL laser using noble gases that also allowed an amazingly low pumping threshold.

Mark Prelas later became Professor of Nuclear Engineering at the University of Missouri where he continued to do pioneering studies of NPLs, including work on unique nuclear pumped "flash lamp" concepts for optical pumping as well as aerosol type NPLs. Fred Boody, who did his MS with me on NPLs, subsequently went to Missouri to obtain a PhD with Mark, and his thesis concerned flash lamp types of NPLs. Mark and Fred's work was very prolific, and as a result they were recognized by the same Russian NPL scientists who set up my visits to Russia, the following photograph shows them arriving at

Arzamas-16 for research discussion and discussion of plans for the next Russian NPL conference which was, for the first time, open to persons from outside of Russia.

*Mark and Fred made several trips to Russia, including one to discuss NPL work with scientists at Arzamas-16. As shown here, a welcoming party met them as they got off a privately operated jet at the air strip used to transport people and supplies to the lab. From the left is the deputy director of Arzamas-16 Anatoly Sinyanskii, Anatoly Zrodnikov, Mark Prelas, Fred Boody, Adreas Ulrich, and Peter Dyachenko is behind Ulrich.*

*This group photo is of the Obninsk NPL Workshop Organizing Committee. Both Mark Prelas and Fred Boody attended to represent the United States. I was invited to the workshop but was unable to travel then due to an important conflict. From left to right first row: Anatoly Zrodnikov, Andreas Ulrich from Germany, Victor Popko, an unidentified scientist, Victoria, Andrey Starostin, and Alexander Pavlovski. Second row: two unidentified senior scientists, M. Troyanov, Fred Boody, Mark Prelas, Peter Dyachenko, and two additional unidentified scientists next to Dyachenko.*

As time went on, I became fascinated with the excimer laser and its possible NPL adaptation. Excimer lasers offer high efficiency and an energy level scheme that seemed well suited for nuclear pumping. However, their power threshold requirement for lasing is very high, making even the high flux obtained in the TRIGA pulse marginal for achieving an NPL version. Still we did proceed to investigate this and obtained considerable data on nuclear-pumped fluorescence of excimers. That in turn led to a series of studies on their use as flash lamps to pump other media. One of my students, Wade Williams, later succeeded in using this technique for NPLs.

Nuclear pumping remains unique in its ability to provide large amounts of input energy (time integrated power) to a laser. Thus, another direction of our research was to seek an energy storage medium to mate with this feature. Singlet delta oxygen, an extremely long-lived metastable state in the oxygen molecule, came to my attention as an obvious choice. This thought was further implanted in my mind when I heard a talk by Alan Garschadin of the Wright Paterson Air Force Base (later to become the Chief Scientist there) about the pumping of singlet delta oxygen with an electrical discharge. He pointed out the main problem was that the average energy of the electrons in an electrical discharge was typically too high, thus favoring undesired ionization over excitation into the singlet delta state. His approach, then, was to find additives that would reduce the average energy of the electrons. Such research has continued until only recently. Several groups, including CU Aerospace Corporation in Champaign, IL, led by David Carroll, finally succeeded in using electrical pumping. The resulting laser was named the e-COIL to distinguish it from the chemically pumped version known as COIL. I know David well and several others in CU Aerospace. Off and on I discussed some aspects of this laser with them, but I did not directly participate in the project.

However, well before that, when I heard Garschadin's desire for a low average energy electron population, I realized that the unique low electron energy distribution from nuclear pumping would make this method for pumping favorable. Thus I quickly mounted a program to study nuclear pumping of singlet delta oxygen for achieving an iodine NPL. This required us to undergo a steep learning curve. Even diagnostic methods for detection of the singlet delta state were difficult. As a start, we built a chemical generation unit to gain practice with production and diagnostics for singlet data. This required chemical substances like hydrogen peroxide and sodium hydroxide, which were tricky to handle. At one point, leakage of chemicals from the singlet delta generator covered the pipes and walls of our lab with a green coating! Meanwhile, the output of singlet delta was led through a long tube across the lab and at some points a red glow developed, signaling successful singlet delta generation. Later, at the reactor, similar phenomena were obtained with nuclear pumping, proving the success of that approach. For a practical laser, however, the singlet delta is mixed with iodine to transfer energy to the laser state in iodine at 1.13 μm. For lasing, the singlet delta state must exceed a certain critical concentration. Mark Zediker, David Shannon, Hani Esayed-Ali, and Cathy Ottinger, thesis students working with me at the time, made great contributions towards development of the required concentration with reactor pumping. Pumping of singlet delta was demonstrated, but actual high-speed flow mixing of iodine with the singlet delta medium was not experimentally done. Instead, we used experimental data to show that the singlet delta levels achieved were indeed adequate to cause lasing in iodine. Later, Wade Williams, in his thesis research, diligently pursued and finally achieved the first iodine NPL using an excimer (X–Br) nuclear-pumped flash lamp instead of singlet delta oxygen.

During the time of our research on NPL pumping of iodine, the chemical version (COIL) was selected by the U.S. Air Force for use in a Boeing 747 as a missile defense system. In fact, however, the efficiency of the NPL-type iodine laser, combined with the unique high energy density offered by a compact pulsed power fission reactor, would offer much improved performance for this airborne laser defense system. But it never gained popularity due to the public perception that use of nuclear reactors in an airplane was dangerous due to possible accidents that might cause the plane to crash in a populated area. Actually, such a reactor would be well sub-critical except when in use for lasing. Various safety analyses confirmed that a crash when "off" would not cause the reactor to go critical.

A related application for NPLs involves space power beaming. As noted earlier, I had vaguely thought about this when I first became involved in NPLs. Now I began looking into the concept more seriously. As scientists at Sandia National Lab and I have pointed out in various papers, there are many advantages in the of nuclear pumping for power beaming, either for land-to-space and space-to-land or for space missions. Multi-megawatt lasers are generally necessary for power beaming, so the facility and power supply would be very large and expensive. A problem encountered with electrically pumped lasers for this use is that the laser is only used periodically when needed and when its path is not blocked by weather, etc. Thus the high cost of the large power supply required for the base is not well utilized. The ability to combine laser pumping with simultaneous electrical power production adds greatly to the versatility and economic viability of the facility. Thus the reactor used to pump the laser would produce commercial electricity while laser energy beaming is running at lower powers or is turned off. This would allow continuous operation of the base nuclear reactor, greatly improving the economics of such systems.

I enjoyed having dedicated and talented students over the years in the NPL laser group, allowing Illinois to pioneer NPL studies over several decades. These students included Russ DeYoung, Jean Guyot, Mark Prelas, Fred Boody, Terry Ganley, Wade Williams, Hani Elsayed-Ali, Yasser Shaban, and Maria Petra. All have gone on to have great careers in a variety of fields. None are currently in NPL research because, as noted earlier, it is no longer being done in the United States. For about five years after leaving Illinois, Russ DeYoung remained heavily involved as the lead scientist for NPL research at the NASA Langley Research Lab. During that period he achieved the highest power research NPL reported up to the time, and also made key advances towards an NPL using uranium hexafluoride medium with the lasing line being in the uranium. As noted earlier, Mark Prelas also continued in the field for some years as Professor at the University of Missouri, Columbia. Among his developments are new concepts for nuclear-pumped flash lamps. After his MS on NPLs at the University of Illinois, Fred Boody joined Mark Prelas at the University of Missouri, where he received his PhD doing flash lamp research under Mark's guidance.

Another challenging aspect of nuclear pumping we tackled in the early work by Jean Guyot and later studies by Maria Petra was how to overcome the so-called "thermal blooming" (or "gas lensing") focusing problem. As nuclear reaction products heat the laser medium, the radial temperature profile across it changes rapidly with time during the pulse. That in turn drastically changes its optical focusing properties. This highly non-linear phenomenon is termed thermal blooming, and it is a real challenge for successful use of a cavity for a NPL because the optics of the light in the cavity then changes dynamically with time. We first realized that thermal blooming was occurring early in our studies when we directed a guide laser beam through the center line of an open (air filled) tube with boron-10 coating its wall. The tube was located in the through port of the TRIGA with the guide laser on one side of the port with its beam passing through the open tube focused on the building wall facing the reactor. When the reactor was pulsed, the light spot on the wall underwent wild gyrations, moving in a spiral-like pattern during the pulse, and then finally relaxing back into its original compact, focused spot. The influence of thermal blooming on laser optics was obvious to anyone observing this visual display. After consulting with my colleague Joe Verdeyen, an expert in laser optics, I decided to use an over-focused cavity design (locating the focal points well behind the reflecting cavity mirrors) such that the gyrations did not completely detune the cavity. This expeditious solution was not without penalty. The gain was reduced from an optimally tuned cavity, increasing the pumping threshold, which perhaps prevented discovery of some high threshold lasers. Still, it worked well for many laser studies, and was used through much of the Illinois work. Later, researchers at Sandia labs used a more elaborate self-tuning design for their cavities. Also, the Russian studies found some ingenious solutions that were not openly discussed. Realizing the importance of the problem, Maria Petra undertook extensive research on this problem for her thesis. She found some additional possible methods to handle the effect, including use of combined wall and volume pumping to reduce the temperature variations.

As discussed earlier, my association with the Russian NPL community was a fascinating aspect of NPL research, and it stands out as a major event in my life. It was spread out over much of the time I worked in this area and even extends on to today. Yet, it happened so gradually that I initially failed to recognize the full significance of this interaction. While attending a European plasma physics conference in Moscow, a Russian scientist, Alex Filyukov, approached me and struck up a conversation about my NPL work. This was during the height of the Cold War, so these conversations were often vague and guarded on their part, though I had nothing to hide. Alex's questions made it clear that he studied all of my papers on the subject in great detail — he even knew some details that I had forgotten! He implied that I was on "the right track," but that he and Russian colleagues had some great concepts they could not yet disclose. Later, Alex invited my wife, Liz, and I to his apartment for dinner. As noted in the earlier photo caption, he was clearly high up in the communist establishment. His wife, Luba, lived with him, but also had a separate apartment where she restored ancient Russian icons under government contract. The restored icons were intended for placement in state-owned museums. Alex was from Georgia. The meal he served us was, as he said, "authentic Georgian," with wild fowl meat, "grass" salad and lots of vodka. I restricted my vodka intake to a single drink, and Liz somehow managed to refuse it altogether, much to Alex's dismay. We especially admired two items in his house, an old samovar (pictured in the opening of this chapter) and some elk horns. Despite our vigorous protests, we left with them as souvenirs to take home. Getting the gifts out through the Russian airport inspection might be difficult. Thus, Alex said he would drive us to the airport the next day to help get them through. However, the vodka got to him (he no doubt enjoyed more after we left his apartment) and he engaged an assistant from his lab to escort us. The assistant flashed some identification cards and got us by the normal security line in order to avoid questions about the samovar and the elk horns (by this time, Liz had the horns strapped around her waist). The remaining guard at the gate glanced at the samovar box and then at the elk horns on Liz. He put hands up to his head and did a little "elk" dance as he ushered us into the gate to board the Aeroflot plane home.

Despite my jovial description, there were some tense moments for us during visits to Russia. We clearly had several KGB agents following us at all times. The hotel rooms were bugged and suitcases were inspected when we went out of the hotel room. In general, all of that surveillance left us, and other visitors in those days, somewhat uneasy. Thus, the passengers on the departing flight, including Liz and I, let out a "cheer" of relief as the airplane finally lifted wheels from the Moscow airport runway.

Following that meeting, I received several letters (clearly passed through Soviet censors) from various Russian scientists working on NPLs. These letters were coordinated through Vladimir Danilychev, one of Nikolay Basov's right-hand people at the famous Lebedev Physical Institute (a key laser lab) in Moscow. Nikolay Basov, a Nobel prize winner for the first neutron emission experiments in Inertial Confinement Fusion (ICF) laser research, was head of the lab, and had also served as Mayor of Moscow. Clearly, Nikolay was not only an outstanding scientist, but an important member of the communist party. Vladimir was working on ion beam pumping of excimer lasers, but kept alluding to a connection with NPLs. He invited me to take several trips to Russia, including an offer to cover all local expenses. He proposed a seminar on nuclear pumping at Moscow State University, group discussions at the Lebedev Physical Institute, and added lectures and discussions at major labs in Siberia and St. Petersburg. I accepted the invitations, but tried to combine things into one circular trip. All visits on this "tour" turned out to be quite unusual. For example, we arrived early for my lecture at Moscow State University, and I immediately was ushered into the president's office where I was served fine food and drink while several others escorted Liz on a tour of Moscow. One of the hosts requested my slides so they could set up the projection equipment. After lunch, we sat around talking for several hours. Then, it was announced that my seminar slides had been reviewed and were considered "classified," so it was cancelled! This made absolutely no sense. How could my talk be classified by the Russians? Was it infighting between various labs and university groups, or was it simply bureaucratic bumbling? No one would discuss it, and to this day I don't know what caused those strange actions. I should add that such actions/confusion were not uncommon in Russia in those days, and had been experienced by other scientific visitors.

Later, when the Soviet Union collapsed, I found out that the University and Lebedev efforts were supporting the major classified NPL projects at Obninsk, Chelyabinsk-70, and Arzamas-16. Indeed, with the new openness, I was invited to visit the formerly secret labs and discuss NPLs with their scientists. I did, and as it happens, I was among the first half dozen foreigners to visit Arzamas-16 and Chelyabinsk-70 (the "Los Alamos" and "Livermore" of Russia). The other visitors were high-level state department officials except for one NPL scientist, Mark Prelas. (Interestingly, aspects of my visit parallel those described by Siegfried Hecker, former director of LANL, in "Adventures in Scientific Nuclear Diplomacy," published in *Physics Today* and listed in the references for this chapter.) The NPL effort was clearly a major project at these secret labs. This actually was centered at around a half dozen research reactors specially designed for NPL experiments. The work force was large, probably exceeding five hundred scientists and technicians. Some of the leading Soviet laser and nuclear scientists were involved, including famed nuclear bomb designer Yulii Borisovich Khariton. All papers and work on NPLs had been classified. Consequently, this work and the high level staff involved escaped recognition by the outside world. The Russians remained confused and did not believe me when I explained that research in the United States was small in comparison. Some classified studies were located at the Sandia National Laboratories, plus independent work like mine occurred at several universities. I maintained (though it was not fully accepted) that my work was independent of the Sandia effort, which was, like the Russian program, classified. Indeed, it became obvious that the keen interest in my work in previous years had been an attempt to gain insight into what they perceived to be a large classified U.S. program. Because I was openly talking about NPLs, it seemed to them like a good route to gain insight into the classified U.S. work.

In addition to the technical insights gained from my visits to the former secret Russian science cities, I also learned about the life scientists led in these remote, secluded sites. It was in many ways great compared to the Soviet life elsewhere. Scientists were very highly valued in Russian society, and these sites had been designed to provide them with many advantages that were not even available even in big cities like Moscow. I visited both elementary and high school science classes plus university-level schools and found superb facilities, energetic teachers, and excited students. I visited a ballet and modern dance academy for youth and found it to be the same. There was a first-rate library, museum, shopping center, and performing arts center. In other words, this was a very well designed city structure which encouraged arts, learning, and athletics, along with a comfortable living style. Additionally, it offered short commute times to work. At the time I visited, however, a "black cloud" hung over the two cities. With the end of the Cold War, the Russian government was in turmoil and it appeared that the long-standing government financial support for many of the nonessential facilities, like the beautiful center for performing arts, would be drastically cut. The director of this center was told that he should start earning money as was done in "capitalistic societies." The director didn't have any concept of how to do that and had never written a grant proposal. The communist government had simply supplied them with a nice budget every year. Indeed, he asked me how our Krannert Center for the Performing Arts at the University of Illinois did such things.

I got him in touch with our director, but lost touch with the discussion as it carried on for months. I would like to visit again someday to see how all of this has worked out. My impression is that the center and such facilities are still there, but that they have significantly changed in character. For example, there are probably much fewer visiting artists from outside compared to the days when the central government brought great artists from throughout the Soviet Union there to perform at government expense. Of course, the actual science labs themselves also faced reduced support levels, but I think they have muddled through all of this turmoil and remain strong. This was partly due to the amazingly quick recovery of the Russian economy, largely because of their sale of oil and natural gas to European countries, and unfortunately the entrepreneurship of organized crime.

I was also curious about the status of religion in the cities, and I learned from discussions with several scientists of a large ancient Orthodox cathedral about 20 miles south of Chelyabinsk-70. After I mentioned this to my hosts, they arranged for a visit. The next morning an armored car with four armed soldiers showed up at my hotel, along with several motorcycle-mounted troops. We took off for the cathedral at high speed with the motorcycles leading the way down a narrow highway. Our vehicle ran smack down the middle of the road with blinking lights and a siren blasting away. We actually forced several horse-drawn wagons off the road, nearly into a ditch. I seemed to be the only one who felt guilty about this. At the cathedral, I was given time to talk with several priests. During the communist regime, religion had been banned, so this cathedral was closed to everyone but the priests living there. However, I learned that their services had actually continued by going underground. Now, with the new government politics coming in, they were repairing damage and fixing up the beautiful and massive cathedral for a revival of its previous glory days of packed worship services.

When discussing religion under the communists, I must add another memory that my wife and I have frequently recalled from an earlier visit to Novosibirsk, Siberia. My wife, Liz, was assigned a young lady who was fluent in English from their scientific translation office to escort her around while I went into the labs. On one occasion, they took a walk on a path through a beautiful wooded area. The subject of religion came up. The guide said that she did read the Bible, but that she had to use one in her grandmother's house because bibles could not be checked out from libraries. As she was saying this, both she and Liz heard a rustling in the nearby brush. The guide immediately said, somewhat loudly, "Of course, I really did not care to read the Bible." It went without saying that she feared ears in the woods. Indeed, throughout our visits, both Liz and I realized there were ears and eyes on us, sometimes blatantly.

As mentioned earlier, I played an indirect role in initiating the U.S. classified NPL program. This happened after I had been working on NPLs almost 6 years, through an unexpected Black Swan-like event. I was a regular attendee at an annual quantum electrodynamics meeting, chaired by Marlan Scully (from Texas A&M University). This meeting was traditionally held in January at the Snowbird Ski Resort in Utah. It was arranged with allotted time for skiing between morning and evening sessions. (I was not a great skier but enjoyed the mountains and the "chickadee" hill area for beginners. Several times I ventured higher only to find myself purposely falling to slow down on the steep slopes.) At one of these meetings, I had the "honor" of giving the last talk on the last day of the meeting. This talk began at about 10:00 PM. This was not a desirable position to be in. Some people had already left; most attendees were fighting to keep their eyes open. A few failed, as was evident by an occasional snore. A hard day on the ski hills was the obvious culprit. I bravely rambled through my talk on new concepts for nuclear pumping of excimer lasers, not really noticing that one participant, Victor George of LLNL, was carefully listening. Earlier he had been one of the pioneers in excimer laser development at LLNL. He became fascinated as he quickly grasped the potential for depositing high amounts of energy into the laser medium by nuclear pumping with a compact pulsed fast reactor. Victor was well placed in the upper levels at LLNL. Upon returning, he gained support to create a classified NPL program. He knew some personnel at a pulsed reactor at the Idaho National Test Site in Idaho Falls, and he obtained support to utilize it for this newly created NPL research project. Victor called me and asked if I would be a consultant to the project, and I gladly accepted. While I did not particularly like it being a classified project, my love of NPLs drummed up my enthusiasm. Subsequently, I made a number of consulting visits to Idaho using the Q-clearance that I then held at LLNL from my consulting there over the years on fusion. As it turns out, the project was too ambitious, and the Idaho staff did not have the background to pull it off quickly as Victor had hoped. Meanwhile, Edward Teller at LLNL drained resources from the NPL work to put into an x-ray laser project, one of his personal pet directions that he lobbied for in congress. Consequently, Victor brought Sandia Lab into the effort to help it survive. They had a history of NPL work extending back to David McArthur's pioneering achievement on the CO NPL. Also, Sandia had several pulsed nuclear reactors that are well suited for NPL experiments. Idaho eventually bowed out of the project, leaving Sandia as the home for the U.S. classified NPL project. Sandia had a dedicated staff, but a limited budget compared to the expense involved in such work. That was the situation in the United States as I visited the former secret labs in Russia.

Sandia workers also learned of the Russian effort, but could not obtain details. Even if invited, their participation in a classified project prevented them from taking trips to Russia or openly discussing their work. After the end of the Soviet Union they made a consorted effort through payment for reports to learn as much as possible about the Russian NPL work. This was not unusual, and in fact encouraged by our State Department as one way to get funding to nuclear experts in Russia to help keep them from leaving to work for "unfriendly" countries that were attempting to establish nuclear weapons programs. They did incorporate some information gained from the Russians into their effort, but it is not clear if they, or I, ever learned all there was (or is) to know about the Russian effort. Indeed, as I noted earlier, it still continues at a reasonable level today. Several years ago I discovered a book on NPLs had been published in Russia. It was authored by Sergey P. Melnikov, Anatolii A. Sinyanskiy, and Alexandr N. Sizov. Several students have helped me translate this book into English, and I have an agreement with the Russian authors to add some material about our work and publish the English version soon through Springer Scientific Publications. This book will provide more insight into the extensive Russian research, but it still omits some classified aspects.

The Sandia program met a different fate. It was building up to a major demonstration when funding was cut. NPLs were not viewed as an important part of our defense program, and possible civilian users would require a large nuclear reactor, which was then out of favor. Ron Lipinski at Sandia (a University of Illinois graduate) made a good case to NASA that power beaming from Earth, from spaceships, or from a lunar base would be an extremely valuable asset in future space exploration. However, NASA too was "pulling its horns in" and viewed this as too futuristic. Thus, today I don't know of a NPL project in the United States. Recently, a Japanese student visited me to discuss the subject since he was doing a PhD thesis on it. Also, there appeared to be several other university-level efforts in China, but the larger Russian effort remains unique.

During the Cold War days I remained in contact with Vladimir Danilychev, who hosted several of my visits to Russia. Toward the end of the Cold War, he gained more freedom to travel outside of the Soviet Union in the company of other Soviet scientists. He led several Russian delegations to the International Society for Optics and Photonics (SPIE) laser meetings. Clearly he was a trusted communist party member who was expected to keep the delegation in line and prevent divulgence of any Russian secrets. Just to be sure, several obvious KGB agents accompanied the delegation, all of whom were housed in adjoining rooms in the conference hotel. The purpose of all of this seemed to be an attempt to showcase the Russian work but not betray classified aspects. Vladimir always kindly invited me up to their room for a drink of Russian vodka and conversation. During the second such meeting in New Orleans at the Hyatt hotel, we held our usual private meeting in their hotel room. Vladimir did not seem at ease, but I did not probe why. I told him I had to leave the meeting a day early to return home due to obligations there. He said goodbye and "stay in touch." Thus, I was somewhat astounded when he called me at my home two days later and announced that he had defected. He was somewhere in the southern United States, but would not say where. When the delegation arrived at the New Orleans Airport to depart, Vladimir slipped away at the last moment, leaving the others to board without knowing what had happened. I never learned the details. Apparently, our CIA had aided him in this daring plan in exchange for some of his knowledge of the Russian laser effort. I heard that they did that sort of thing for key people. Vladimir did not contact me again for several years, as he was in hiding to prevent any possible repercussions. I found he was living in Irvine, CA, where he works for a small high tech company that grew out of work with the University of California, Irvine. Later, when I attended a professional meeting in the area, I visited him and we reminisced about the old times. I have never tried to pry into any of his private affairs with Russia or the U.S. government officials involved in helping him defect. He did volunteer to say that he had been assigned to learn all he could about the U.S. NPL program from me during my visits to Russia.

As time went on I was fortunate to have a series of three young NPL Russian scientists (Evgeniy Poletaev, Andrei Fedenev, and Gregory Batyrbekov), each from a different lab, come to the United States to work with me for periods of 6 months to 2 years. Due to the sensitivity of the subject, clearing all of the paperwork for these visits was a feat within itself, but exceedingly worthwhile. They enjoyed the accessibility of the TRIGA reactor with its through beam port for NPL experiments, and the good availability for time on the reactor. They also really enjoyed being here in America, and they were fun to be around. In addition to scientific exchanges, I learned much about varied aspects of life in Russia from them. We all still correspond off and on and fondly remember those extended visits. I also remain in close contact with Anatoly Zrodnikov, current head of the lab at Obninisk. Anatoly was an upcoming young NPL scientist when I first met him.

The era of my NPL research ended in the late 1980s. A final chapter was later closed when the University of Illinois decided to close the TRIGA reactor after 40 years of service.

## Further Reading


S. Hecker, "Adventures in Scientific Nuclear Diplomacy," *Physics Today*, July, pp. 31–37 (2011).

S. P. Melnikov, A. A. Sinyanskiy, A. N. Sizov, G. H. Miley, *Lasers with Nuclear Pumping*, New York: Springer Business and Science Media (in press).

G. H. Miley, D. McArthur, M. Prelas and R. DeYoung, "Fission Reactor Pumped Lasers: History and Prospects," *Proceedings, ANS Conf. on Fifty Years with Nuclear Fission*, NIST, eds. J. W. Behrens and A. D. Carlson La Grange Park, IL, pp. 333–342, Apr. (1989).

G. H. Miley, "Nuclear Pumping of the Iodine Laser — Revisited," *Proceedings of the Santa Fe High-Power Laser Ablation Conference* (SFHPLAC), ed. by Claude R. Phipps, *SPIE Proceedings Series*, Bellingham, Washington, Vol. 3343, pp. 692–700 (1998).

G. H. Miley, "A Nuclear Pumped Flashlamp Iodine Laser," *Proceedings*, U.S.–Japan Seminar on Physics of High Power Laser Matter Interaction, Kyoto, Japan (1992).

G. H. Miley, E. Batyrbekov, E. Suzuki, M. Petra Y. Shaban, E. Poletaev, "Concepts for a Hybrid Fission Electric NPL Power Plant," *7th Intern. Conf. on Emerging Nuclear Energy Systems* (ICENCES '93), ed. Hideshi Yasuda, World Scientific, Singapore, pp. 367–371 (1994).

G. H. Miley and M. Petra, "Thermal Blooming in NPLs," *Proceedings of SPIE International Symposium on High-Power Laser Ablation*, Santa Fe, NM, Apr. 26–30 (1998).

# Chapter 5

# Direct Electron Beam Pumped Laser

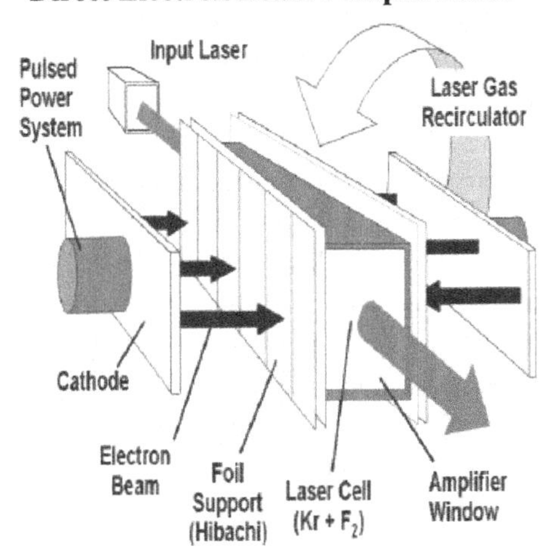


*Key components of a modern electron beam pumped KrF amplifier employed at the Naval Research Lab in Washington, DC, for Inertial Confinement Fusion Experiments. My earlier e-beam pumped laser experiment at Cornell was not so elegant, but used the same basic principles. It represented the first achievement of a direct electron beam pumped laser. Had I not returned to Illinois and NPL research shortly after the experiment at Cornell, I might well have been at the forefront of this very rapidly growing field — a Black Swan sighting missed!*

In 1969, Norman Rostoker (then Head of the Applied Physics Department at Cornell University) invited me to Cornell for a year to assist in studies of relativistic electron beam propagation. The objective of this project was the acceleration of ions by electrostatic drag forces in the electron beam thought to create GeV ions. However, shortly after I arrived, it was found that with the beam conditions possible at that time, this mechanism would not work. Thus, I was left to explore my own interests. I began by teaching a course on fusion technology, which was continued as a formal course offering after I left Cornell. I also began theoretical research studies on field reversal for magnetic confinement using the relativistic electron beam. This work was pioneered there by Ravi Sudan and Hans Fleishmann. In the process it dawned on me that, by analogy to nuclear pumping, the electron beam would be an excellent means of directly pumping lasers. The beam could deposit a large power and energy density into a gaseous laser medium, thus "pumping" laser excited states. I devised a method of doing this and reported the achievement of an electron beam pumped laser in early 1971. This was the first report of directly pumping a laser using an electron emitting diode. A "diode" in this context is simply composed of two parallel conducting plates separated by a vacuum region. One plate was electrically grounded while the other was pulsed to megavolt (MeV) potentials of positive sign. Electrons were extracted from the grounded plate by the high electric field, and then passed through the thin outer plate of mesh, entering the laser medium where the pumping action occurred. My design used a curved section of the beam guide with mirrors on the ends to form an optical cavity as done for conventional lasers. However, I soon discovered that the power density was so high that in selected gases like nitrogen, the medium was "super radiant," forming a laser without the need for mirrors as predicted theoretically in a pioneering paper by Robert Dicke of Princeton University. The closest prior work was a traveling wave laser pumping study at the Naval Research Lab. Indeed, shortly after my announcement, the interest in direct diode pumping of lasers virtually "exploded" in both research and commercial labs. Unfortunately, I left Cornell soon after this success to return to Illinois, where I again picked up on nuclear pumping research. Thus, my initiation of this field rapidly faded from view as others took over the lead to advance e-beam pumped laser technology. Because I left the field without capitalizing on my success with the e-beam laser, what might have been a Black Swan faded into a Mute Swan in terms of the impact it created.

Interestingly, back at Cornell when the electron beam entered the laser medium and lasing occurred, the laser power levels were so high that a flash of light flooded the whole High Voltage Lab. No one else in the lab worked with lasers then so this light flash was not expected and caused some alarm. Several staff asked that this work be restricted to times when others were not present, such as late at night. Subsequently, the experiments were done after the other lab staff had gone home, by me and a graduate research assistant. The laser and its light flashes did not worry me (but we did wear protective glasses). However, I was not yet fully comfortable with the MeV voltage levels used in the electron beam power supplies. Thus, I was always leery of these devices.

Aside from my research experience, traveling to and living in Ithaca resulted in some interesting times. When we went to Ithaca, I had two cars, a Volkswagen bug and a station wagon. I wanted to get both there so I could drive the VW to work and leave Liz with the station wagon. Fortunately, Bob Twardock, a good friend from the veterinary school, offered to drive the VW up because he wanted to visit a friend there from his days of graduate study at Cornell. Bob and his wife Mary are both tall, so the first challenge was for them to fit into the beetle! They left the week before we did and Bob fondly recounts how even trucks would pass them going up hills (there are plentiful hills en route to Ithaca) and then he would catch up and pass them on the downhill side. We rented a wonderful house from a faculty member going on sabbatical to Europe. It was right on the west side of Cayuga Lake, down a steep dirt road off of Taughannock Falls road. The architect had to have been a student of Frank Lloyd Wright, because the house had the Prairie School style — windows with great views of the lake and glass skylights in the kitchen that were covered with beautiful colored leaves in the fall. When the winter snows arrived I soon realized that it was virtually impossible to get a car back up the dirt road from our house to the highway. Thus I ended up parking the VW by the mail box at the top of the hill, along with a sled. The sled was the fastest way to get down the hill, especially when laden with shopping packages, my briefcase, etc.

*Left: My son Hunter and our dog Bonnie were waiting for me and the sled to arrive at the bottom of the hill in front of our house in Ithaca. Right: The picnic table and slide on our "beach" on Cayuga Lake are barely visible here. The stone steps leading from this beach up to the house are completely covered in this photo, but in the summer were favorite spots for garter snakes to lie on the warm rock in the sun.*

We went to visit my parents in Williamsburg that Christmas, driving the station wagon but leaving the VW in an area on the side of the house. There was a record-setting snow over Christmas, and when we returned it took several days to locate and dig the VW out. We all enjoyed living in Ithaca. We could hear and see sculling boat teams practicing in the lake and passing our house. Our son, though quite young then, was thrilled to find that he could easily catch small garter snakes that liked to warm themselves on the rocks between the house and the lake. He thought they were great for scaring his sister. Our dog, a golden retriever, enjoyed improving her swimming skills in the lake.

We felt somewhat saddened as we packed up to leave the following June. One unexpected event was that Norm Rostoker, the person who invited me to come to Cornell, knocked on our door the morning we were about to set off for Illinois. Norm said excitedly that he had discussed my diode laser work with some people at the Office of Naval Research who were interested in funding it. Thus he proposed that we unpack and stay another year! I was certainly tempted, but was by then fully committed to returning to Illinois.

It is also interesting that I had spoken with Norm several times about my interest in aneutronic fusion which might be implemented by burning p–$^{11}$B in a Field Reversed Configuration (FRC). Norm showed little interest at the time. Later Norm moved from Cornell to the University of California, Irvine, where he began studying aneutronic fusion and an ion beam driven version of the FRC. This reversal of interest from his discussion with me is not surprising. Norm was characteristically slow to show enthusiasm for something until he had investigated it further. This lead to the formation of a company named Tri-Alpha (based on p–$^{11}$B fusion which produces 3 alpha particles as the reaction products). This company is funded by Paul Allen, a founding member of Microsoft. At the time of this book, Tri-Alpha is the largest privately funded fusion company in the United States.

The time and my diode-pumped laser research at Cornell passed quickly, but I left with many new friends at Cornell. Because my wife, Liz, is a Cornell graduate, I have seen some of them again off and on when I accompany her to class reunions. I had become especially good friends with Hans Fleischmann and Ravi Sudan. Both were highly respected in physics research. Thus, Ravi's death a few years ago following some years of illness was a sad event. I had visited him by his bedside earlier during one of Liz's class reunions. Ravi was internationally known for his studies of plasma instabilities and I am indebted to him for adding much to my understanding in the field. Hans Fleischman also provided me with new insights into experimental technique. He has an especially quick ability to make rough "order of magnitude" calculations in his head. I soon learned not to argue about Hans' quick "estimates" unless I had carefully done the full computation in advance.

In closing, I will note an interesting event that happened several years before my time at Cornell. Liz returned to Cornell alone for a week to attend a "retreat" arranged by the alumni office. This included some lectures and discussions led by various faculty. Hans Bethe was one of these speakers, and he ate with the group that evening. Liz sat near him and at one point he said to her with a smile, "Tell George he is wasting his time on fusion!" I was not surprised, though this was meant somewhat as a joke. Despite his Nobel Prize for work on the nuclear reactions which supply the energy in stars, Hans is known to strongly support fast breeder reactors as a "best" future power source. Hopefully Hans was wrong about fusion, but unless we find a way to speed up its development I might begrudgingly agree with him.

## Further Reading

G. H. Miley, "Relativistic Electron Beam Excitation of Gas Lasers," *Intern. Symp. on Electron, Ion, and Laser Beam Tech.*, San Francisco Press, pp. 279–290, May (1971).

G. H. Miley, "Research and Teaching Fusion in Nuclear Engineering," *J. of Eng. Ed.*, pp. 1–10 (1971).

# Chapter 6

# Advanced Lasers

*I attended a U.S.–Japan NSF laser meeting in Japan in the 1980s where I spoke about nuclear pumped lasers. In this photograph of the participants, I'm shown kneeling (fifth person over from the left in the first row). Mark Prelas, now a professor at the University of Missouri–Columbia, is in the dark coat in the front row. He did his thesis with me on NPLs and has gone on to a very successful career with research ranging from carbon nanostructures to radiation-induced chemical reactions and fusion. He and I have continued to interact off and on, especially relative to nuclear pumped laser work by the Russians and also on nuclear batteries. The first four in the front row (from left) are Sadao Nakai, then head of the laser fusion lab at Osaka, Chiyoe Yamanaka, founder of that lab, and Art Guenther, the Chief Scientist at Kirkland Air Force Base — all good friends and collaborators.*

In addition to studying NPLs, my students and I performed some forefront studies of other types of lasers. Joe Herzig did a thesis on a unique thermionic driven He–Ne laser. It was designed to be incorporated in the fuel element of a fission reactor. This concept was along the lines of using thermionic conversion of heat to electricity in fission reactors, studied for many years at General Atomics (GA) and by Ned Rasor's thermionics consulting company. Interestingly, the head of the thermionic reactor group at GA was my old colleague, Dick Dahlberg, from KAPL. Also, Ned Rasor and I knew each other from professional meetings and often discussed direct energy conversion using a variety of methods. Joe Herzig and I envisioned a dual laser–electric fission reactor. His experimental laser was very successful, but our discussions with Dick Dahlberg and Ned Rasor about combining concepts did not gain momentum. I think they felt that their hands were full with the thermionic-electric reactor work, much less adding lasing to the technology. Indeed, they were probably right, because high temperature materials and radiation effects challenge the thermionic-electric reactor development such that the project gradually disappeared. However, I would not be surprised to see it come back some day, and indeed, with a laser included. Instead of "co-generation," this would be a first "tri-generation," providing electricity, heat, and laser light!

In concurrent work, students and I sought other applications of nuclear pumping that involved fusion. A future vision for fusion was to incorporate NPLs in an ICF facility. This was designed with the NPL laser medium surrounding the fusion chamber to form a blanket-like region that served to moderate and absorb the D–T fusion neutrons. A portion served the normal blanket heat–tritium recovery functions. The remainder of the blanket region served to use the neutrons from the fusion implosion reaction to pump multiple NPLs. With proper optical delay time, the NPL lasing would occur after the reaction chamber was cleared and a new target was injected. This concept then resulted in a nuclear-pumped "feedback-laser driver." With high fusion gain targets, the lasers only consumed part of the neutrons produced, the other fraction going to tritium breeding and for heat production in the blanket for conversion to electricity as done in a normal ICF power plant. The details for this feedback laser system were worked out in a design study, but with fusion power itself still some years off, the concept remains a distant vision.

Several other concepts for other types of advanced lasers were considered at this time. In one, ions from a small accelerator were employed to pump gaseous xenon medium producing an intense VUV/UV flash lamp. Such a flash lamp was intended to serve as a next-step light source for the semiconductor computer chip industry. While some interest by people in that field was shown, this approach never gained much traction compared to other laser and Z-pinch sources under consideration.

Another entirely different direction in laser work was inspired by studies in Russia done by Alexander Karabut. He reported an amazing soft x-ray laser based on bombardment of a metal target such as palladium with pulsed deuterium plasma. Karabut was a close friend of Andrei Lipson. Andrei was a leading solid state physicist in Russia. He worked with me for four years as a visiting professor, so I obtained some firsthand insights into Karabut's x-ray laser experiments through Andrei.

Thus I was pleased to see that Karabut planned to present a paper at the 11th International Conference on Emerging Nuclear Energy Systems where I was also scheduled to present a paper. This meeting was in Albuquerque, NM in 2002 and turned out to be quite eventful. Another person who attended, Bob Smith, was a friend of mine who I had met at various cold fusion (ICCF) meetings. Bob was retired from the Air Force, and in addition to being a fighter pilot he had a strong nuclear background. He had attended the Air Force Institute of Technology in Ohio to obtain a degree in Nuclear Engineering and had been sent to ORSAT, the nuclear reactor school set up at Oak Ridge National Laboratory to train attendees about the basics of nuclear reactors. Upon retirement, Bob set up a consulting company in Washington, DC and among other things became involved in cold fusion. When I went to the Albuquerque meeting I ran into Bob, and while talking I mentioned Karabut's unique x-ray laser. I told him I was thinking of using a pulsed discharge in my IEC device (see Chapter 11) to bombard a target like Karabut had done in order to duplicate his amazing experiment. Bob took immediate interest and suggested that we should try to get support to duplicate the experiment from the office of joint military services technology located in Albuquerque. Such a laser would seem to be of great interest to this office, which was supporting R&D on several other advanced lasers. Bob said he knew several of the staff there and would set up a meeting. Thus both of us were ready to listen to Karabut's talk carefully and hoped to have him go with us to the joint services office. This plan hit a real snag when we discovered that Karabut's English was almost impossible to understand. Neither Bob nor I spoke Russian, so this was a real barrier. Again, by accident, we ran into Haiko Lietz, a retired high school physics teacher whose parents were from Russia. Thus he spoke fluent Russian. Haiko then helped us discuss the experiment with Karabut and also generously agreed to accompany us to the joint-services office to serve as an interpreter.

After a series of discussions, a complicated proposal, and other negotiations, I ended up with a Phase I grant from the joint services office to do a soft laser experiment. Bob Smith was a consultant on the project. Duplication of the pulsed plasma discharge conditions used by Karabut was difficult. This was especially complicated because we had listed Karabut as a consultant on the project with funding for him to spend two months with me during the set-up of the experiment. Now that the project was confirmed, Karabut wanted additional funding, because he argued that my IEC set-up was not adequate for the discharge and proposed to bring his set-up with him — for a price. The money he wanted in order to do that would exceed the total contract budget, and was therefore impossible. Clearly Karabut did not understand this, but wanted to take full advantage of his invention. Meanwhile, his difficulty reading and responding to our emails (written in English) added to the problem. In addition, his laser claims were not fully documented with traditional data like threshold and gain measurements. Hence some questions remained about his claims. As a result, Phase I of the project had modest funding intended to clear up these issues. If successful, Phase II would be much more aggressive and could better cover Karabut's demands. Faced with this dilemma, I decided to go ahead without Karabut's direct assistance. As noted earlier, I already knew more details about the experiment than appeared in the published paper from discussions with Andrei Lipson, who had observed the experiment. My experiment ran into a number of difficulties related to achieving a uniform discharge under the extreme pulse conditions. We finally managed to obtain strong soft x-ray emission, but did not achieve a laser. To fully explore that possibility, we decided a larger power source was needed to further increase the discharge intensity.

Our proposal for Phase II to do that did not receive funding, however. In addition to increasing concern about claims of lasing, questions about practical uses for the soft x-rays came up. The point is that the soft x-ray penetration depth in materials or even air is small. This can be understood by noting that their energy, about 2 keV, is far below that used in dental x-ray machines (~ 80 keV). The fact that intense x-rays were obtained at all in the experiment was (and still is) very exciting because their generation under these discharge conditions requires very non-linear physics effects. But, without further support I was not able to continue this work, so the experiment was shut down. In any case, this had been an interesting and exciting experience.

One side benefit of this episode was that Haiko Leitz, who helped with the Russian translation at the original meeting, became very interested in this laser and the related field of cold fusion. Afterward he began attending ICCF meetings and has become heavily involved in Condensed Matter Nuclear Science (CMNS) discussion groups.

More recently, my interest returned to the singlet delta oxygen–iodine laser. However, in this case, instead of nuclear pumping, I proposed a unique "wire cage" electrical discharge design to pump the oxygen. The discharge configuration was inspired by the wire cage Z-pinch studies at Sandia Laboratory that uses a huge Marx capacitor bank discharges to produce Terawatt electrical impulses through the wire cage that then undergoes a magnetic pinch action. This creates megawatts of x-ray power from the metallic plasma that is formed in the implosion of the cage. In my version, the system is miniaturized and most importantly, arc channels in the gaseous median form the cage structure (versus actual metal wires). This approach allows a high repetition rate because the cage is quickly reformed by the discharge itself after each pulse (versus changing out wire cages, as is done at Sandia). Another key feature is that the electron energy distribution has a soft spectrum, such as desired for efficient singlet delta pumping. This idea generated considerable interest when I presented it in a talk at a 2009 American Physical Society meeting on plasma physics. But the studies of it are only in their infancy, and many unanswered questions remain. Because this work is unfunded, I had to put it on the "back burner," compared to my other active research.

## Further Reading

G. H. Miley, "Small Ion Driven VUV/UV Light Sources for Laboratory Applications," *Proceedings of SPIE*, Vol. 5196, p. 12, Aug. 4–6 (2003).

G. H. Miley, A. G. Lipson, N. Luo, and A. B. Karabut, "On a Photon-Driven Solid-State X-ray Laser," *Inertial Fusion Sciences Applications*, Editors: B. A. Hammel, et al., pp. 1074–1076 (2003).

G. H. Miley, "Controlled Filament Non-Local Discharge (CFND) for Laser Pumping," *Proceedings of 92nd Annual Meeting of Optical Society of America*, Abstract JSuA22, October 19–24 (2008).

# Chapter 7

# Alpha Particle Effects in Thermonuclear Fusion Devices

*Winnie Kernbichler (center skier), Professor at the University of Graz, Austria, hosted this winter workshop on fusion in the Austrian Alps. Winnie had spent some time at Illinois with me and gained great enthusiasm about alpha particle effects and advanced fuel fusion. Thus he invited me to lecture on those topics at the workshop. As seen here, we enjoyed skiing in addition to the meeting. This meeting was one of those planned in Austria to help get them into the European Union fusion grant program, an objective that ended up being successful.*

*In my early work on alpha particle behavior in fusion systems, I collaborated with people at the Culham Laboratory in the UK, where the Joint European Tokamak (JET) was under construction. My role was to help in the design of instrumentation to measure alpha particle effects. I'm shown here in front of a model of JET during one of my visits. The JET was eventually completed and very successfully operated, and it provided forefront alpha particle measurements such as the energy and angle of high energy alphas escaping to the wall. Due to a "sunset clause" in the original European Union contract to support JET, it should have been shut down several decades ago. However, it has continued to run with the objective of providing improved data for actual D–T fusion plasmas, being one of the few facilities in the world capable of such operation. (Other large Tokamaks typically use deuterium discharges to avoid tritium complications.) The comparable experiment in the United States was the Tokamak Fusion Test Reactor (TFTR) at the Princeton Plasma Physics*

*Laboratory, and it also provided forefront D–T data. Unfortunately, it was shut down some time ago due to funding crunches. It was argued, incorrectly, that most of the data that could be obtained from it had already been obtained. That decision is just one of the many unfortunate mistakes that have been made in the U.S. fusion program.*

Early on in my studies of fusion systems, I wanted to identify a key area to concentrate on, because a well-defined focus enables better progress and provides more visibility than do random studies. Some criteria I used were that the subject was not yet well explored, and that it should involve basic fusion plasma physics. Burning fusion plasma physics is fundamental to the successful operation of a fusion reactor and something that I had previous experience with. Thus I identified alpha particle effects in a fusion reactor as an issue for my students and me to attack. At that time most of the rest of the fusion community was devoted to improving plasma confinement in Tokamak experiments using non-fusing hydrogen plasmas, not on future reactor-grade plasmas. This approach was adopted because most of the community thought we were "20 years off from a fusion reactor" so improving plasma confinement was the issue receiving most study. This left the important area of alpha particle effects wide open for new work.

In my later fusion work I have focused on advanced fuel fusion, e.g., D–$^{3}$He and p–$^{11}$B. However, I was initially focused on D–T fusion, as was the target of the worldwide fusion community. Indeed, plans were being formulated to build two major D–T burning Tokamaks to study alpha particle effects — the Joint European Torus (JET) in England and the Tokamak Fusion Test Reactor (TFTR) at Princeton University in the United States. When D–T fuses, the reaction releases a 3.5-MeV alpha particle plus a 14-MeV neutron. The neutron escapes the plasma and enters the surrounding blanket to provide a means of breeding tritium and heating, which is used in a thermal cycle to produce electricity. However, the electronically charged alpha particle remains confined by the magnetic field in a typical magnetic fusion reactor. Thus, it slows down to equilibrium with the other ions in the plasma at around 20 keV. This "thermalization" process deposits the kinetic energy of the alpha particle in the plasma such that external heat sources are not required (i.e., the burning plasma is said to be "ignited"). While a reactor could operate sub-ignited with some external heating, ignition is generally preferred in order to obtain a high fusion energy gain. Once the alpha particles (an ionized helium atom) are thermalized in the plasma, they can be thought of as being "ash" from the reaction. That is, the alpha particles do not fuse, but still "dilute" the plasma, reducing the D–T ion density, and hence the fusion rate.

There are some key issues about this process that had to be better understood, however. First, a fraction of the high energy alphas may have large-sized orbits in the confining magnetic field of the device, allowing them to swing out of the plasma and hit the wall. Associated with this is the possibility that the alpha is born on a trajectory which has an angle relative to the magnetic field such that it is not confined by the field. Such an alpha particle is then said to lie in a "loss cone," defined by a conical region in velocity space that bounds unconfined orbits. The second problem is that a high energy fusion alpha particle has an initial velocity above a velocity termed the "Alfven velocity" characteristic of so-called "Alfven waves" in many of the Tokamak-type confinement systems. Thus, there is a possibility that as the alpha slows down and it hits a speed equal to the Alfven velocity, its energy may be efficiently transferred to the Alfven wave, and thus lost relative to heating the background plasma particles. This is generally termed the "Alfven instability." In fact, there may be several more subtle instabilities that occur where energy is lost to other plasma waves. A third problem is that the thermalized alpha ash may not diffuse out of the plasma rapidly enough. Its density can build to the point where the plasma burn is quenched. If the ash diffusion occurs at basically the same rate as the other plasma ion species, this is not a problem, and the alpha concentration will remain manageable. However, there are some subtle effects, such as electric-field-driven drifts, which might slow alpha diffusion out of the plasma. Plus, the alpha particles and the main plasma particles are ultimately taken out of the system through a "diverter" plate and a pump. Alpha particles (helium) do not pump as well as the other species, so special care must be taken in this part of the system to prevent the reduced pumping speed from causing accumulation of "ash" near the diverter. This would ultimately back ash up into the total system. In our work on alpha particle physics, my students and I investigated all of these issues. Indeed, a number of students, including Lee Hively, Meka Papa Rao, Wesley Downum, Ming-Yang Wang, Bill Sutton, Glen Sager, Tom Petrie, Junpeng Guo, Shu Kay Ho, Shin Hu, and Ken Wesley did their theses on several of these topics. This represented a sizeable effort when compared to other groups working on burning plasmas. We were early in this area of study, so many of our investigations were path-finding. However, as both JET and TFTR progressed, many more plasma scientists worldwide undertook related studies, in some cases building on the original work done by us.

The diagnostic group at JET, headed by Neal Jarvis, designed a unique "wall detector" to measure the number and direction of loss cone alphas and alphas undergoing the unconfined orbital drifts hitting the wall. This detector was designed using data from our calculations, so I made several trips to the United Kingdom to discuss this detector concept with the JET diagnostics group. As it turned out, getting to D–T burns in both TFTR and JET took longer than originally hoped. The delays were largely due to logistic concerns about handling the radioactive tritium in these experiment systems. Once JET became operational with D–T plasmas, extensive and incredibly valuable experimental data on alpha particle losses to the wall were obtained with the JET wall detector system. The Alfven wave instability remained a worry, but scientists working on JET and TFTR expanded the original studies to carefully incorporate realistic geometry and magnetic field structure characteristic of these devices into the calculation. Then, following a series of important papers on that subject, they identified, at least theoretically, windows of operation where the instability could be avoided. These calculations were backed up with simulated alpha particle experiments where high energy ions were injected into non-burning hydrogen plasma to cover a range of conditions passing through and out of the operation window.

While TFTR and JET underwent a number of very successful D–T burn experiments, the burn time was always too short to experimentally study alpha ash buildup effects. Some other supporting data has been obtained where helium has been purposely injected into the plasma. Based on all of this data, the International Thermonuclear Experimental

Reactor (ITER), the next development in Tokamaks, is now under construction in France is expected to avoid burn quench by alpha "ash" buildup. Indeed, one of the main physics objectives of ITER is to study long burn time effects such as this. I, however, remain a "hold out." Some of our analysis has suggested that electric field effects in the plasma might cause higher ash buildup than expected, thus reducing the fusion burn rate and causing a problem. I have repeated this position at several meetings, but the general consensus is that this view is against the "common wisdom" of the ITER community. Unfortunately, the projected time for the first D–T experiments in ITER keeps moving out into the future. (The first hydrogen plasma in ITER is scheduled for 2020, but D–T operation would not begin until 2027.) I can only hope to be around to find out what really happens.

One of my students, Shin Hu, did a thesis on the issue of alpha "ash" buildup. She was a very bright student but somewhat reserved with a soft voice. Thus I was quite surprised about an event that occurred when Shin and I drove to Argonne National Lab one day to discuss computation techniques with researchers there. After leaving an interstate, we were making our way through one of several suburbs that had speed limits as low as 20 mph in places. I was aware that this area was known for its speed traps, and I was driving carefully when I suddenly heard a police siren behind me. The officer asked if I realized I was speeding. I said I didn't think I was, and he said something about writing a ticket when Hu suddenly said loudly and defiantly, "Professor Miley was not over the speed limit. I was watching carefully!" I ended up with only a warning. Shin was very forceful when she wanted to be!

Shin Hu graduated with a great enthusiasm about fusion, but to get a job back in her home country Taiwan she joined their national fission program in a reactor safety group. We have exchanged Christmas cards for some years. In these cards she frequently talked about her daughter and son, who she hoped would follow in her footsteps and go into physics or nuclear engineering. As such, I was not completely surprised when several years ago her daughter applied to enter graduate school in my department at the University of Illinois. She also applied to Purdue, and ended up going there due to an excellent assistanceship offer. Later she showed up at my office one day to introduce herself. I was so pleased, plus I was surprised to learn that she was on campus to participate in a badminton match between Illinois and Purdue. After she left, I sat back thinking, "It is so great to see a new generation carrying forward."

Our studies of alpha particle effects in magnetic confinement systems slowly wound down as interest in this area grew because of ITER, and the government labs threw many resources and people at the problem. We could not compete with these larger organized efforts. However, Inertial Confinement Fusion (ICF), or "laser fusion," was then still in an early stage relative to equivalent alpha particle studies. Thus our interest gradually turned to studies of "plasma burn physics" in ICF, as described in Chapter 10.

## Further Reading

S. Hu, "Thermal Helium Buildup and Control in a Fusing Tokamak," Ph.D. Thesis, Nuclear Engineering, University of Illinois (1992).

G. H. Miley and L. M. Hively, "Superthermal Alpha Transport and Wall Bombardment in Tokamak," *J. Nucl. Matls.*, Vol. 76, pp. 389–395 (1978).

G. H. Miley, "On Alpha Heating in Toroidal Devices," *Fusion Technology*, Pergamon Press, pp. 133–139 (1979).

G. H. Miley and C. K. Choi, "Effects of Alpha Drift and Instabilities on Tokamak Plasma Edge Conditions," *J. Nucl. Matls.*, Vol. 121, pp. 92–99 (1984).

G. H. Miley, "Some Aspects of Alpha Particle Transport and Kinetic Instabilities in Tokamaks and Mirrors," *Physica Scripta*, Vol. T16, pp. 58–71 (1987).

G. H. Miley, S. Hu and V. Varadarajan, "Epithermal-Thermal Alpha Transport and Control," *Fusion Technology*, Vol. 18, p. 633 (1990).

# Chapter 8

# Alternate Fusion Concepts

*Left: Liz and I are shown with Heinz Hora (left) and Manfred Heindler (right) at an advanced fuel fusion meeting in the Netherlands. As discussed previously, I've had numerous technical collaborations with Heinz over the years. I have also collaborated with Manfred on advanced fusion concepts. Manfred was a Professor of Physics at Graz University at the time of this photograph. Later, he went on to become the Minister of Energy for Austria. We first met when he and Winnie Kernbichler, Assistant Professor at Graz, hosted an international meeting on "Emerging Nuclear Energy" at the University. After that, I continued to interact with both on advanced fuel fusion, a topic that I introduced to them at the meeting. Before that, little fusion research was done at Graz. Manfred and Winnie decided that advanced fuel fusion was a way to "leapfrog" into the field because virtually no studies of this were ongoing elsewhere in Europe. Right: In the late 1970s, Manfred Heindler and his wife, Claudia, spent a semester at the University of Illinois while he was on sabbatical from Graz University. He came to collaborate on advanced fuel fusion studies. During his visit, I found that I had to spend a month at LLNL working on an experiment I had initiated. Liz went with me to Livermore, and the Heindlers "baby-sat" our house. Among other things, they took care of our West Highland White dog which they fondly referred to as the "white flash." Later, as described in the text, Manfred purchased an old Buick station wagon to drive across the country, and this turned out to be a real "adventure" with many car breakdowns. Sadly, he passed away of cancer in the late 1990s.*

When I decided to create a strong fusion research effort at Illinois, I faced the obstacle of finding a focus that would distinguish this program. As noted in the previous chapter, work on alpha particle effects provided one focus, but I also sought more comprehensive system type areas. This was complicated because strong fusion programs already existed in other universities in the Midwest and on the coasts. Some had elaborate magnetic confinement experimental laboratories in the Physics or Electrical Engineering. However, at Illinois solid state physics had been so strong that plasmas had never gained much of a foothold outside of a few specialized labs like the Gaseous Electronics Laboratory where my friend Joe Verdeyen had some fusion research.

After considering several possibilities, I chose the concept of a "driven" mirror fusion reactor designed to produce neutrons for fusion materials testing. This concept was an outgrowth of several converging factors in my experience. I had already spent several summers at LLNL, where I worked on direct energy conversion concepts for mirror reactors. In those days LLNL was the main center for mirror research, so I was immersed in that thinking. My time at LLNL in the summers between 1962 and 1970 were both enjoyable and a time of professional growth for me. Ralph Moir, a "fixture" at LLNL working on the mirror fusion program, invited me to work there the first summer and served as my mentor (and remained so for many subsequent visits). He and his colleague, Bill Barr, were the leaders in direct energy conversion concepts and experiments (as well as other areas like fusion-fission hybrids, liquid metal blankets, blanket material issues, and tritium breeding). I came in that summer enthusiastic about tackling everything. Ralph said "slow down — one step at a time!" That year, I ended up studying some problems in high-voltage holding under conditions encountered in direct energy conversion devices and neutral-beam injectors. Fortunately I was able to make some advances in that problem and published a well-cited paper on the subject. Ralph was right — it's best to do a good job on well-defined problems and then move forward. Ralph also taught me about the "culture" of LLNL relative to its fusion and weapons sides. He introduced me to many people, some of whom I later collaborated with or learned from. Ralph is still a good friend, but we don't see each other as much anymore. In later years, I continued as a consultant to the mirror program and also gained a clearance to work on ICF target physics (a highly classified area in those days).

*Left: During my sabbatical stay in Japan in the 1990s, I was visiting professor at Nagoya University giving lectures on advanced fuel fusion. I also spent time at the nearby National Institute of Fusion Science (NIFS) facility shown here. NIFS was nearing completion and*

*housed the experiment "large helicon" which is a type of stellarator. This experiment, sponsored by the university community in Japan, was viewed as potentially leading to use of advanced fuel fusion, although D–T is a first objective. As a university facility, NIFS is quite independent from the government JAERI laboratories (the equivalent of our DOE labs). The JAERI laboratories are in northern Japan and focused on Tokamaks, plus ITER support. Right: I'm shown here at NIFS along with Hiromu Momota, my host and long term colleague (third from left). Hiromu is flanked by section leaders for plasma physics and technology at NIFS.*

One experience I distinctly remember was when Ken Fowler, then Head of the Mirror Division at LLNL, called me and several other consultants to a meeting to help decide what to do as DOE funding for standard mirrors was suddenly cut from the budget. (This budget cut was claimed to be due to the overall fusion budget being reduced, and the fragile position mirrors were in was due to their need to recirculate energy at extremely high efficiency to sustain a net energy production. Some critics argued the hope that direct energy conversion could enable net energy was too optimistic. No doubt there was also other politics involved in this decision, but such inner workings have never been disclosed.) After much debate, in answer to Ken Fowler's request I recommended converting the division's effort to the Field Reversed Mirror (FRM). Others favored the Tandem Mirror, which was the final selection. The Tandem Mirror, like the Tokamak, was originally proposed by the Russians. It used a clever means of electrostatically "plugging" the mirror loss cone from leakage in order to obtain a much improved fusion energy gain. If it worked as projected, direct energy conversion would not be required for successful operation. Also, its physics was thought to be better understood than the FRM. However, the FRM offered a relatively small size plus a combination of a closed internal magnetic field giving good confinement, but also a natural diverter region. Because of these ideal features, I maintained that it would be a superior reactor if its physics worked out according to theory. I guess I have always been willing to gamble on the underdog. Anyway, a large Tandem Mirror experiment was constructed at LLNL over the next few years. But, due to politics, its funding was cancelled just as the first plasma was ushered in at a gala ceremony attended by many dignitaries including several congressmen and high-level DOE staff. What a waste of taxpayers' money! It should have been cancelled much earlier, or, after spending so much money, it should have at least been run long enough for a fair evaluation. I still wonder what would have happened had my choice of the FRM been chosen instead. If a smaller device like the FRM had been selected, perhaps experiments could have started earlier, before the DOE budget issue that plagued the Tandem Mirror became so serious.

Another possible route that was discussed in addition to the FRM was to develop a conventional mirror as a neutron source, not a power reactor. The fusion community was desperate (and still is) to have a facility to study effects of the high energy 14.1-MeV fusion neutron on materials. Various accelerator driven neutron source concepts had been proposed. However these designs usually had either too small a volume of high flux to allow reasonable size samples, or failed to offer sufficient fluence (time integrated flux) to fully satisfy needs for damage studies. Both problems could be overcome by using a driven mirror system (one where a net electrical input is supplied to drive the sub-breakeven fusion reactor). The mirror seemed a logical choice because the sample volume and fluence would be adequate. In addition, a steady-state, but low gain, mirror reactor seemed to be close at hand based on existing experiments at LLNL. While this approach had appeal, as I already noted, the Tandem Mirror for electrical production was selected as the next focus for the LLNL program.

Indeed, this discussion brings to mind the very beginning of my work on fusion systems at the University of Illinois. My first experience with fusion reactor design per se came indirectly when I was asked to go to the University of Wisconsin as part of a DOE panel to review a proposal to carry out a Tokamak design study, named UWMAK. Professor Harold Forsen, the Principal Investigator, had put together an interdisciplinary team to carry out this study, ranging from plasma physics scientists and nuclear engineers on to materials science and structural engineers. He said this would be a "path-finding" study for fusion development, and stressed that the University had the unique ability to pull together a diverse team of experts to cover the wide range of topics involved. These arguments were very convincing and I voted with the majority of the review team in favor of awarding the contract. They got it! That started off a long-standing tradition of fusion reactor studies at the University of Wisconsin. Harold Forsen later left for new challenges at Bechtel, but young professors like Gerry Kulcinski and Bob Conn assumed the leadership and went on to great careers.

After I returned home, I kept thinking, "I should be able to do that too!" However, as I thought about it, I wanted to work on a more near-term concept. My experience at LLNL brought to mind mirror fusion. It is simpler to build than a Wisconsin Tokamak, and hence it seemed more realistic for an early reactor. There was a major problem, however. As already noted, the leakage through the open loss cones formed in the magnetic "bottle" field prevented a decent fusion gain (energy out/in). The thought came to me, "Why not run a driven mirror?"(i.e., supply more electrical power in than comes out). The fusing plasma would still be great for physics studies and for an intense 14.1 MeV neutron source if run on D–T fuel. Later, that could be the basis for a fusion–fission hybrid reactor, but even more pressing at the time (and perhaps even now) was the ability to use the neutrons for studies of neutron effects on materials of interest for a fusion reactor. I reasoned that, if the mirror was not run too far below energy breakeven, people would be quite willing to pay for the net electrical power input. I first put together a small faculty planning group to work on this, to help raise money for a design study and also to find people with the diverse backgrounds needed for the design team. In a university, faculty are usually focused on their research and not easily convinced to make time to work on an interdisciplinary project like this. However, this project was so exciting that I found I could get some key people without much trouble. In searching for money, I took the bold leap of calling a vice president at Illinois Power, our local electrical utility company. They typically did not invest in research like this, but after several long phone conversations and a visit, I convinced officials at Illinois Power to support this project. It seemed reasonably near-term and would lead to early development of fusion, which could greatly impact Illinois Power if they were on the ground floor.

The grant I received from Illinois Power was not as large as the DOE UWMAK grant to the University of Wisconsin. However, it did the job. The driven mirror neutron source to be designed was termed "Illiball." Top staff from various university departments joined the effort and enjoyed learning from each other. All of this caught the LLNL people off-guard. As home of the mirror, they did not want to admit others had gotten ahead in any aspect of mirror technology. So, they implied this was an old thought to them, and then later started a study of mirror neutron source facility of their own. When published, our conceptual design was well received by both the broader fusion community and the utilities as represented by their research arm, the Electric Power Research Institute (EPRI). My goal, my Black Swan, was to perform some preparatory pilot unit type experiments and then build the Illiball. But the timing was wrong. Neither DOE nor the utilities were ready to do that for a variety of reasons, ranging from a then-declining fusion budget in DOE to arguments about the best approach. (Other possibilities included accelerator-target neutron sources or doing nothing, and relying instead on the theory of neutron damage.) So after the end of the Illiball project, the collaborating faculty and I went in other directions. It might be noted that the fusion community agrees (some more strongly than others) that a fusion neutron source must be built soon for materials damage studies. Japan was to be given money to do precisely that when they lost the hard-fought battle to have the ITER site in Japan (the site went to France). However, due to slow funding, the international neutron source facility remains years behind schedule.

I continued to view mirror systems, particularly the FRM, as a possible route to fusion to overcome the problems I perceived with Tokamaks. It was during one of my visits to the United Kingdom to attend a meeting at the Culham Science Centre (the major UK fusion lab) that my early worries about the wisdom of Tokamak type fusion grew. During a reception prior to the meeting, Bob Caruthers (a long time scientist at Culham, who performed many fusion reaction design studies over those early years) invited me and others to go downstairs for a drink of some special scotch and to play a game. The scotch was delightful (Bob was a true Scot and a connoisseur of fine scotch). The game was a version of "pin the tail on the donkey." However, in this version, the blindfolded contestant was asked to pin a light water fission reactor (a paper cutout) on the drawing of a Tokamak posted on a tavern wall. After some wild stabs by various contestants, Bob changed the rules so we could peek out beneath the blindfolds. The contestants managed to pin 8 or 9 light water reactors squarely within the outline of the Tokamak. "How," Bob asked, "could a Tokamak ever compete as a power reactor in view of its low overall power density?"

When I returned home, Charlie Baker (another "game" participant from the nearby Argonne National Laboratory) and I wrote a letter to Ed Kintner, then Head of the Office of Fusion Energy for the U.S. DOE, expressing that worry. Ed Kintner had formerly been high up in the Nuclear Navy program. Thus, he understood fission reactors and immediately grasped the dilemma. However, he did not despair of Tokamaks. Instead, he instigated new studies to identify methods to further improve their performance. Charlie Baker went on to lead many Tokamak studies, always seeking solutions and never losing faith. On the other hand, I turned my interest even more strongly to alternate confinement concepts, thinking that I could find better solutions there. This eventually led to studies of alternate magnetic confinement configurations including Field Reversed Configurations (FRC), Field Reversed Pinches (FRP), "advanced" Inertial Confinement Fusion (ICF), and Inertial Electrostatic Confinement (IEC; my work on the latter two continues today and will be discussed later). It might be noted that soon after that meeting Bob Caruthers stopped doing design studies and retired, perhaps partially in frustration over this predicament. His "pin the light water fission reactor on the Tokamak" game must have been a way for him to vent this long pent-up frustration. Plus, he wanted to warn other younger fusion scientists of the situation in a memorable way.

My approach to studies of alternative fusion confinement concepts was to seek out one or two of the key issues in their plasma physics and tackle them first. This was intended to build a realistic physics background as a basis for the reactor design. Others using extensive computer simulations in national labs and universities were working on such problems, but I usually sought shortcuts based on physical assumptions to do quick "first-order" modeling. For example, in studies of the FRC, one of my early students, Ed Morse (now Professor of Nuclear Engineering at UC Berkley), investigated use of the Hill's vortex fluid model to gain a simple representation of the equilibrium for the FRC. I presented this concept in a seminar at LLNL, but the theorists there were not enthusiastic. They thought the issue of stabilization of the bad magnetic line curvature was not adequately addressed. Hill's vortex uses ideal hydrodynamics without consideration of particle kinetic effects which provide the stabilization. I argued that large orbit effects provided the stabilization and in turn allowed formation of a Hill's-vortex-like equilibrium. Thus I felt that the Hill's vortex analogy provided a good approximation for the global plasma behavior. This then allowed rapid evaluation of many key features of the FRC without lengthy and sometimes frustrating plasma simulations. That in turn provided a reasonably realistic basis for FRC reactor studies. Later, Ed Morse did a detailed study for his thesis using the LLNL supercomputer (at that time, one of the few supercomputers available for plasma studies and one of the largest in the United States) to perform the complex plasma kinetic computation required to study FRC equilibrium. His work bore out the reasonable validity of the Hill's vortex assumption. Meanwhile, my students and I had applied this assumption to a series of FRC studies, including both a FRC power reactor and a FRC fusion space propulsion study. The assumption allowed such system designs to proceed rapidly, which put us ahead of other groups who did not venture into this area due to uncertainty about the equilibrium and how to analyze it.

My introduction to field reversal and the FRC originally came from participation in an early IAEA fusion meeting in Culham, England. One of the presentations there by Dan Wells of Miami University discussed the theory of "plasmoids" (small plasma "balls"), which he had formed experimentally and successfully translated over long distances, even colliding them together in a series of pioneering experiments. In effect, he discussed the theory of stabilization of plasmoids by forming a configuration that minimized the total electromagnetic and plasma energies. However, he presented the math in such an obscure manner that the audience generally failed to appreciate its significance until much later.

The next paper was by Brian Taylor, head of the Culham fusion laboratory theory group. I found his talk spellbinding. Taylor discussed the famous ZETA fusion experiment performed earlier at Culham. This high current pinch experiment had been touted by Culham scientists in a headline newspaper release as achieving strong D–D neutron emission, suggesting that breakeven conditions could be reached if the United Kingdom nuclear research funding agency would provide the money for a somewhat larger device. Unfortunately, some months later, it was found that the neutrons observed were not from a bulk plasma reaction, but from plasma instabilities throwing the plasma against the wall where it interacted with deuterium absorbed on the surface. That type of fusion would not scale up to a power reactor which requires reactions throughout the plasma volume! Funding for ZETA was dropped due to this error and the subsequent public embarrassment it caused. Culham Scientists learned a lesson: check carefully before submitting news releases. The scientific approach is to publish a peer-reviewed paper first before making public claims. (Too bad they did not pass this wisdom on to Martin Fleischmann and Stanley Pons, who caused a "ZETA"-like effect over the public announcement of "cold fusion," discussed later in the chapter on LENR.)

At this IAEA meeting, some ten years after the ZETA experiment was closed down, Taylor dramatically unrolled a strip chart from some old ZETA runs and pointed out how the plasma initially went through a series of rapid instabilities. Then, it settled into a quiescent stable phase. He pointed out that this phase was always accompanied by characteristic changes in the magnetic field probes. Based on this observation, Taylor produced an elegant and relatively simple (and understandable) theory showing how the characteristic magnetic field changes provide a minimum energy state. This elegant theory caught much attention and has continued to be widely used. As it turns out, Dan Wells' theory was very similar and also provided important insight into minimum energy stabilization by field reversal processes. Unfortunately, as I already noted, Wells' complex presentation failed to catch much attention and it remained obscure for some years before people realized this. Taylor's ability to produce simple elegant treatments of complex phenomena is something I try to emulate. It requires understanding of crucial aspects in a problem so the simple treatment of that issue provides good accuracy. Such insight only comes with study and experience.

My subsequent venture into the plasma focus, one of the older alternate fusion confinement concepts, had some interesting and also bumpy times. I had read about the focus and was fascinated with the theories for its pinch effect, which were varied and still contested. I came up with one which I presented at an IEEE plasma science meeting in Buffalo, NY. While waiting for my luggage at the Buffalo airport, I accidentally came across Frank Mead, a well-known scientist at Edwards Air Force Base recognized for his advanced propulsion concepts. We shared a taxi from the airport to the meeting and got to know each other better. In the conversation I told Frank about my interest in the possibility of burning D–$^3$He in a plasma focus. (D–$^3$He is one of the "advanced fusion fuels" that minimizes neutron production, i.e., induced radioactivity in the reactor structure, as well as radioactive tritium versus D–T fuel.) The lack of magnetic field in the focus would eliminate cyclotron radiation losses and, according to my new theory, the focus confinement was adequate for D–$^3$He fusion in it to approach breakeven. Due to its relatively simple structure and natural formation of exhaust plasma to provide thrust, the focus has been viewed for some years as a possible route to space propulsion. If it could use D–$^3$He, problems of neutron induced material damage and complicated tritium breeding would be avoided. Thus I brought up my thoughts to Frank. As it turned out, Frank had already thought about the plasma focus for space propulsion and liked my idea. This initiated a long association, ending several years ago when Frank retired from Edwards Air Force Base to live at a remote ranch in the mountains about 60 miles from Edwards.

I have a few comments to provide the flavor of this episode in my research. I obtained a 10 kJ capacitor tank from LANL and used that to build a small plasma focus experiment. A variety of students and colleagues were involved in the focus and experiments on it. Chan Choi, who came from a PhD and post doctoral position at Southern Illinois University to work with me, was extremely active. Later, when he accepted a faculty position at Purdue, our joint collaboration with Frank Mead continued. Glenn Gerdin, along with Tom Blue and John Gilligan (all young and enthusiastic research staff I had hired to work on fusion) took over active leadership in the focus work. Francisco Verneri, a graduate assistant in the lab, had begun work on energetic ion emission during the pinch phase of the plasma focus, and Glenn Gerdin assumed responsibility for guiding Francisco's thesis, which concerned this effect.

Meanwhile, I learned more about the intricacies of the focus, largely from attending a fusion-focused workshop at Steven's Institute of Technology, just across the river from downtown New York City. Frank Mead insisted that I should attend. He had heard about this meeting and even managed to cover my travel expenses through an AF contract. This was the home of pioneering focus studies by Winston Bostick, Jan Brzosko, and colleagues. They had studied the formation of small high density hot plasma formed by collapsing electron "filaments" during the focus pinch. The claim was that the majority of fusion reactions occurred in these filaments where, due to the very high local density, fusion reactions such as burning advanced fuels could occur. They viewed this as the pathway to a fusion power reactor. The trick was to create conditions to reliably form these filaments and to measure their properties with precise diagnostics. They were making progress. I learned much from them during the workshop and we became friends. The problem was, however, that I had been advocating formation of a large volume fusion region in the compressed focus plasma, not filaments. They opposed that, and in my defense, I claimed that filament collapse was really an unwanted instability that would not scale up in power. The resolution of these views did not occur, but one result of all of this discussion was that I convinced Frank Mead that I must move to a larger plasma focus to prove that our approach could scale in power.

*Chan Choi is shown here with me during my retirement celebration. Chan and I began collaboration on alpha particle effects, and we continued on to work on the dense plasma focus and ICF, including the AFLINT target concept where the D–T core is ignited and burn propagates into the surrounding deuterium. We remain in close contact since he moved to Purdue University, only a few hours' drive from Champaign.*

I obtained a surplus capacitor bank (several MegaJoules) from the Betatron project at Illinois, which had been shut down due to funding problems. The Betatron accelerator was invented at the University of Illinois, and its fascinating history is a whole story within itself. The inventor, Donald Kerst, later moved to the University of Wisconsin. He was "in love" with this capacitor bank. When he heard I had acquired it, he called several times to see if it could be moved to Wisconsin. I was certainly reluctant to do that, but felt bad not to honor Kerst's request after all of his work on it. However, it turned out that he could not raise the sizeable cost needed to move the bank. Meanwhile, I obtained an AF contract through Frank Mead to convert it to a fast modern Marx bank configuration suitable for plasma focus work. That required development of state-of-the-art fast current switches and low capacitance "blumline" circuitry. Fortunately, Mike Williams, a senior laboratory technician in the Fusion Studies Lab, had the ability to design, build, and make things like that work. Olivier Barnouin, a PhD student from the Omega Fusion Laser Laboratory at the University of Rochester, also joined the focus group as a postdoc. Olivier provided the boost needed to really move ahead rapidly. Not only was he experienced, but he enthusiastically devoted many long hours of work on the construction of the new capacitor bank and then to experiments on the focus. Later Olivier returned to France, his original home. The plasma focus experiment produced some important research papers. But considerable money would be needed for advanced experiments with a larger unit. Meanwhile I wanted to consider other approaches.

Because Illinois did not have a Tokamak experiment (which dominated most magnetic fusion research worldwide), I was initially drawn to study alternative confinement concepts to develop a niche position in the field. This also complemented my interest in advanced fuels like D–$^{3}$He and p–$^{11}$B. An alternate confinement scheme with a low internal magnetic field inside of the plasma (so-called "high $\beta$" confinement) plus high ion-to-electron temperature (a non-Maxwellian equilibrium) is essential for burning advanced fuels. By this time a small but growing number of fusion scientists were becoming concerned that Larry Lidsky's article designating the D–T Tokamak as a "white elephant" (not a desirable power plant, even if it worked) might prove to be correct. These issues provided a strong motivation to consider alternate confinement concepts, and as it turned out, over the years many of my students were involved in that research. Thus the use of advanced fuels in a variety of concepts such as the RFPs, the FRC, Spheromaks, multiple mirrors, etc., were all covered in thesis studies. Students involved were very enthusiastic and creative, giving the Fusion Studies Laboratory at the University of Illinois prominence. The students involved included: Ed Morse, John Galambos, Phil Stroud, Mark Campbell, Rick Olsen, Dan Driemeyer, Jack DeVeaux, Bill Tettey, Robert Start, Wayne Choe, Ron Miller, M-Y Hsiao, Meng Wang, Chuck Bathke, Ken Werley, Q. T. Fang, Francisco Verneri, Yang Yang, and Rick Nebel. (Technically, the IEC is also an alternate confinement concept. But due to its special history at the University of Illinois and with me, it is treated as a separate chapter and students involved are noted there. Otherwise, this list would be even longer.) Many have remained in fusion and gained great reputations. Some have gone into other fields and also done well. Each has been a special part of my life and I could write pages about that, but don't have space here.

Another event that had a lasting influence on my thinking about alternate fusion concepts came from a call from Starnes Walker, a senior physicist at Phillips Petroleum in Bartlesville, OK. Starnes asked me to consult on a fusion project they were planning which was yet another alternate confinement scheme. He said Phillips' management was willing to undertake the high financial risk of a fusion experiment because, like a "wild cat" oil well, the payoff could be tremendous. They wanted to work on an alternate confinement scheme, not so much to allow alternate fuels, but to achieve a smaller fusion power plant than allowed by Tokamaks. I agreed to consult. It was exciting to have a private company seriously involved in fusion. Starnes had a PhD in Physics from UC Berkley and was very experienced in various experimental radiation diagnostics. The plan was to join with General Atomics (GA) in San Diego to develop the OHTE (Ohmically Heated Thermonuclear Experiment) fusion device, a modified toroidal reversed field pinch configuration with an internal diverter proposed by Tehiro Ohkawa, a senior VP at GA. GA was also working on a DOE sponsored Tokamak project known as DIII-D. Thus, GA management understood fusion. In contrast, there were not many scientists at the Phillips Laboratory knowledgeable in fusion plasma physics. But Starnes quickly gathered a small crew of quite knowledgeable staff and consultants. As the project continued he gradually added more staff, including Jack DeVeaux, one of the students who assisted me at Illinois on this project. Unfortunately another knowledgeable Phillips scientist with a Nuclear Engineering background, Mike Driscoll, left to become Professor of Nuclear Engineering at MIT just before Starnes started the fusion project. Mike had been a classmate of mine at Carnegie Tech and remained a good friend. In fact, I had visited the Phillips research laboratories at Bartlesville several times to give lectures at Mike's invitation. It would have been fun to work with Mike on a fusion project.

Administrators at GA were initially concerned that starting a privately funded experiment, even though modest in size compared to DOE's Tokamak experiment there, called DIII-D, might cause ill will at DOE. The concern was that DOE might perceive that this new project pulled some top people off of DIII-D. The decision to go ahead was facilitated by the fact that the Gulf Oil Corporation had recently taken over GA, so two "sister" oil companies, Gulf and Phillips, would be associated in the project. Some very experienced GA staff members were assigned to the OHTE project. Starnes' plan was for Phillips to provide financial support, support diagnostics, and perform analysis of experimental data. Starnes rented a house in La Jolla near GA and, along with others from Phillips, spent extended periods of time there. He had the consultants, including myself, come out for project reviews once every few months. Tom Dolan, then a Professor at University of Missouri, Rolla, also became a consultant. I knew Tom well staring from his days as a student at Illinois where he did his thesis in the Gaseous Electronic Lab on IEC physics. Following his retirement from Missouri, he is now an Adjunct Faculty member of NPRE at Illinois. Thus I enjoyed the chance to work with Tom on OHTE.

These meetings and the project continued with good progress for several years. Then, several events occurred that might be termed as "negative" Black Swans for commercially-supported fusion research. First, T. Boone Pickens, a famous corporate "raider", threatened both Phillips Petroleum and Gulf Oil. Gulf Oil had sold much of the property they owned surrounding GA to capitalize on the very high land prices in that part of La

Jolla. That income and other large profits made them a good takeover target. After a struggle, Gulf Oil collapsed and was no more! This is particularly unbelievable to someone like me who, having been raised in the Pittsburgh area, knew the skyscraper Gulf Oil headquarters building there with its landmark weather forecasting lights on top. Plus, as already described, one summer while a student at Carnegie Tech I had worked in the world famous Gulf Research Laboratory in Pittsburgh. Meanwhile, Phillips Petroleum moved quickly to sell off assets and build a debt to become unattractive for takeover. One cost-cutting target was advanced research like the OHTE fusion project. Starnes learned that if some other "outside matching" money could be obtained, OHTE might still continue. Progress on OHTE had been good so Starnes proposed to DOE that they move the project to Los Alamos National Laboratory (LANL) to take advantage of a large pulsed power supply available there. I am not sure of all the details, but despite this opportunity to involve both Phillips and GA in the national fusion program via OHTE, DOE declined. Part of the objection was that this was a non-Tokamak program. With Gulf Oil gone, GA is now owned by the Blue brothers from Canada who have focused on work on defense contracts. However the DIII-D Tokamak continues with DOE support as one of the largest fusion experiments in the United States.

Starnes later joined a consulting firm in Cleveland with a major responsibility for the design of the DOE's supercollider project in Texas. After that project was cancelled by congress, Starnes eventually ended up in Washington, DC, as Chief Scientist for the Office of Naval Research. Later he moved to a similar position in the Office of Homeland Security. Our paths cross every year or so at meetings. This whole episode was fascinating, but due to its unfortunate end, it represents a sad part of my fusion experience. I say "sad" in the sense that bureaucratic organizations in government mixing with capitalistic corporations ended in a stalemate, stopping OHTE. What could have been a positive Black Swan turned out to be a negative one. Another example of this type of problem that immediately comes to mind is when McDonnell-Douglas Corporation proposed to build a scaled-up Elmo Bumpy Torus (EBT) fusion experiment at ORNL. That is a long, involved story that I don't have space to go into, but the result was again negative. DOE ended up turning that proposal down.

It's interesting how this early research evolved into other relationships and work. My interest in alternative configurations to burn D–$^3$He and fusion space propulsion led me to meet Eric Rice, CEO of ORBITEC Technologies Corp. in Madison, WI. Eric, like my colleague Gerry Kulcinski at the University of Wisconsin (see Chapter 11) is a former football player for the Wisconsin Badgers. Eric tackles technology and his company ORBITEC with vigor and a game plan as though he is preparing for a football game. Thus at one time ORBITEC held a record for the most SBIR contracts granted to any small business nationwide. ORBITEC can move rapidly with a handful of bright young PhDs as employees, and this, along with Eric's leadership, accounts for its productivity. Eric invited me to give a lecture on IEC fusion for space propulsion at a Madison American Institute of Astronautics and Aviation (AIAA) Chapter meeting in 2001. We became good friends afterwards and collaborated on several projects. He insisted that I also join the newly formed AIAA technical committee on space colonization. Eric's interest in this evolved from a project that ORBITEC obtained from NASA aimed at development of methods to create a "green house" for growing food plants on Mars. He had founded this committee and was its driving force. He wanted me to represent the power area relative to space colonization. You don't say "no" to Eric, so I soon found myself as an officer on the committee. I also became the point of contact for fusion power for space missions as well as use of lunar $^3$He resources for D–$^3$He fusion as discussed further in the next chapter.

Any discussion of Black Swans by someone from the University of Illinois must mention Jon Bardeen. Bardeen did not work on plasmas, but his research has touched us all, and is fundamental to the fusion-enabling technology of high-field superconducting magnets. Bardeen is the only person to date to win the Nobel Prize in Physics twice. His awards were for the transistor (which ushered in the age of modern electronics) and for the 1957 BCS (Bardeen, Cooper, and Schrieffer) theory of superconductivity that was so fundamental the effects are still rippling through science and industry. John passed away about 12 years ago, but we overlapped at the University of Illinois for many years. We were not actual collaborators, but knew each other on a first-name basis. He served on several of my students' thesis committees, but my main discussions with him occurred during accidental meetings at a lunch wagon where the two of us sometimes went for a quick sandwich. To really be a close friend of John, one had to be a solid-state physicist or a golfer. John once responded to a reporter's question about what could be better than two Nobel Prizes by saying, "two holes-in-one." John loved golf, and while not looking very athletic had been a record breaking NCAA swimmer while at the University of Wisconsin. *True Genius*, a book about his life (commissioned in his memory by our Physics Department), claims his contributions to science equaled or exceeded Einstein's. But because he was not flamboyant, his accomplishments were not widely recognized by the public. Indeed, John's talks and lectures, while usually deep and thorough, were not delivered emphatically, hence he was called by some "whispering John." For that reason, students did not flock to his classes as much as might have been expected. Yet, I would have given a right arm to have been one of his students. As I discuss in Chapter 18, I feel television and other communications media have unfortunately turned many current students into passive observers wanting to be entertained rather than being interactive learners.

Some of my fusion students, especially international students, wanted to have the honor of having John Bardeen on their PhD committees. I usually tried to discourage this, saying John was not working on plasmas. However, the typical response was that Tokamaks used superconductivity coils! I gave in to several very persistent students and John ended up on a few committees. John would never refuse a request like that from students. One such student (unnamed) was working on fusion reaction charged-product transport and orbits. One day this student appeared in my office saying that he had been working on "a new revolutionary calculus," and wanted to change his thesis to that. I responded that he should transfer to the Math Department to do that. Many discussions ensued. At one point he accused me of trying to suppress science! We eventually compromised with the agreement that he would solve one fusion product orbit calculation this way in an appendix of his thesis. However, that was exceedingly involved despite his claim that his "new" theory simplified calculus. Thus he ended up solving a simple integral equation in the appendix. This same very opinionated student had insisted that John Bardeen be on his committee. John said little during the student's final oral exam presentation. Then, towards the end he spoke up, saying, "seems like a good thesis, but that appendix about integrations simply doesn't belong there." I had told the student that several times before without success. But after Bardeen's objection, it was quickly removed. Unfortunately, I think the student may still blame me for "suppressing" the new calculus. Hopefully he has mellowed over time.

My last discussion with John was just weeks before his death. I mentioned that I thought that cold fusion reactions occurring in a condensed state in solid hydrides involved some physics possibly related to his superconductivity theory. I tried to explain my thinking in more detail and then asked what he thought. He characteristically responded that he "would have to think about that — maybe?" John kept an open mind about cold fusion and scientific questions in general. He was never one to quickly offer an opinion, but when he spoke (or whispered), one should listen carefully. John had achieved two major Black Swan events, and I am sure he would personally claim to have achieved another — he did have several holes-in-one over time playing at the Urbana Country Club golf course! Interestingly, as discussed in Chapter 12, we later showed that the "clusters" we are studying for cold fusion (LENR) are actually a class II superconducting state below ~70°C. Thus there does indeed appear to be some relation to John's theory.

John's approach to science is important to understand. He passed on much of this philosophy to his students. For example, Nick Holonyak, John's first PhD student at the University of Illinois, stayed on in the Electrical Engineering Department. He also has his Black Swan event — the discovery of the Light Emitting Diode (LED). That technology is still revolutionizing large sectors of the industry and our lives. I don't know why he has not yet, as of this writing, received a Nobel Prize, but he has received the Presidential Medal and almost every other award possible. His many students over the years have been leaders in developing LED technology in a variety of industries. In a gala event celebrating his 85th birthday, attendees from across the country recounted many stories about their work with Nick. One story that caught my attention was that over 30 years earlier Nick published an article in *Popular Science Magazine* predicting LED headlights for cars. Some scientists scoffed at the prediction, but in recent years this has begun to occur. Given my optimism about my predictions for fusion, I feel that I can understand Nick's anxiety for quick progress. A book describing Nick's accomplishments is listed in the references for this chapter.

Nick and I used to talk in the men's locker room after exercising at noon in the Kenny Gym (or, as some say, the "Old Men's" Gym) near my office. Until Nick retired, we both frequently exercised during the noon break. Like John Bardeen, Nick did not speculate about scientific issues without solid knowledge about them. However, in many other areas (campus politics, national politics, philosophy of education, life experiences, etc.), he was opinionated and outspoken. His voice carried well, so everyone in the locker room also knew his opinions. One achievement that endeared Nick to all of us was when he went to the President's office to successfully protest the proposed closing of the old gym in favor of a parking lot. Nick had the stature to demand such a meeting and surely had the ability to state his case emphatically. As a user, I think Nick achieved a Mute Swan event for saving the gym!

I did not have the opportunity to collaborate scientifically with Nick or John, as our fields of research did not directly overlap. Still, knowing them has been extremely important to me as outstanding role models for "true scientists." (I use that wording in view of the wonderful book about John Bardeen's life, *True Genius*, mentioned earlier.)

## Further Reading

E. Dummel, *Inventors 101: Spotlight on Nick Holonyak, Including his Education, Most Famous Inventions such as the Light-Emitting Diode, Awards Received, and More,* Webster's Digital Services (2012).

L. Hoddeson, and V. Daitch, *True Genius: The Life and Science of John Bardeen,* Joseph Henry Press (2002).

G. H. Miley, "Compact Tori for Alternate Fuel Fusion," *Nucl. Inst. and Methods*, Vol. 2, pp. 111–120 (1983).

G. H. Miley and J. Nadler, "Technical Opportunities — Subgroup on Emerging Concepts," *Proceedings of the 1999 Snowmass Fusion Summer Study*, Snowmass, CO, Jul. 11–23 (1999).

# Chapter 9

# Advanced Fuel Fusion and Direct Energy Conversion

*Left: I'm shown here with one of my first PhD students, Henry Sampson. Henry did his thesis on direct energy conversion, more specifically the gamma cell. He was also the first African American student to receive a PhD in nuclear engineering in the United States. Interestingly, Henry's father also graduated from the University of Illinois, but that was before my time. Henry went on to have an exciting career at the Aerospace Engineering Corporation in Los Angeles, CA. He also authored the "definitive" text on the history of African Americans in the movie industry. In the last few years, bloggers have mistakenly credited him with the invention of the cell phone. No matter how you look at it, he is an achiever! Right: Yasser Shaban (center) is a fairly recent PhD student who worked on direct energy conversion and nuclear-pumped lasers. His research resulted in arguably the first visible output NPL. He is shown here at an international meeting receiving congratulations for the best poster award at this laser conference. Later, he also worked as a post doctoral associate in the IEC laboratory, after which he returned to Egypt.*

In 1968, I contracted with ERDA (the predecessor to DOE) to write a reference book on fusion energy conversion. This contract came about because of my growing work in fusion and because of the success of my earlier book on *Direct Conversion of Nuclear Energy* (American Nuclear Society, LaGrange, IL, 1973). It was viewed by ERDA as similar to the famous series of early books on nuclear energy, artfully written for them by Samuel Glasstone and co-authors — books that educated many early scientists and engineers on the fundamentals of this new field. (I originally learned nuclear reactor theory from the early reactor theory book by Glasstone and Edlund, and fusion plasma physics from the one on that by Lovberg and Glasstone.) My plan of attack for book writing was to review the topic critically and provide new views, not just rehash available papers. In the process of preparation for the fusion book, it dawned on me that with fusion power being so far off in the future, considerable improvement in energy conversion from fusion plants would be essential to compete with other future energy sources. Thermal cycles of 40% simply seemed shortsighted in light of the ultra-high temperature of the plasma. The route around this dilemma might be direct conversion of the plasma energy to electricity. However, this approach seemed impractical using conventional D–T fusion. In that case only 20% of the fusion energy goes into charge particles (alphas) which could be used for direct conversion. The way around this problem would be to use "advanced fusion fuels" like D–$^{3}$He and p–$^{11}$B where the major fraction of fusion energy released goes into charged particles. Their energy then can be directly converted to other energy forms through coupling with electromagnetic fields using methods like direct electrostatic collection and travelling waves, as I discussed in the *Fusion Energy Conversion* book. The key to this approach revolved around finding an advanced fuel that might be developed in early fusion reactors, possibly leapfrogging the favored D–T approach. D–T offered the largest fusion cross section of any fuel, and was viewed as *the* approach to power by fusion scientists in those days (and continues to be favored by most DOE fusion scientists today). Little was said about other fuels and their cross sections. However, I was aware that Rand McNally at ORNL had done pioneering work identifying cross sections for fusion reactions involving light elements other than D–T through boron. There are a surprising number of reactions, but only a half dozen or so have an adequate cross section magnitude to be of practical interest for a fusion reactor. I carefully reviewed this work with Rand, and discussed possibilities like D–$^{3}$He and p–$^{11}$B in some detail in the fusion book. Plus I changed my own research to move in this direction.

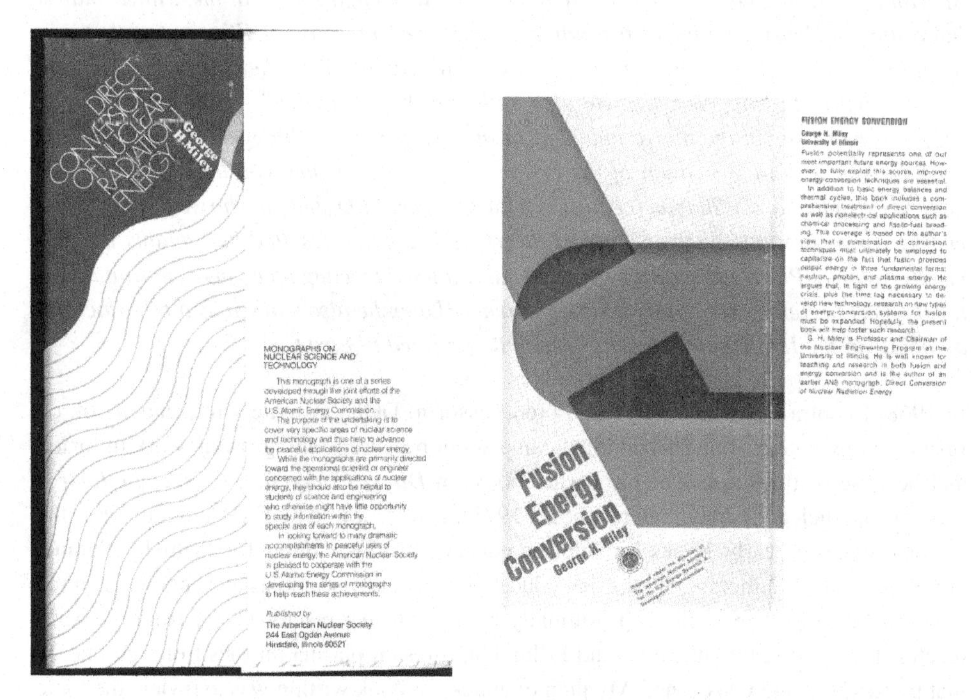


*I authored two books on energy conversion under contract with ERDA (the predecessor to DOE) in the period between 1967–1976. The work on these books led me to formulate*

*personal views on how to go about direct conversion of nuclear energy (from radioisotopes, fission, and fusion sources) to other energy forms, including not only electricity but also forms like optical lasing and chemical production. That has had a profound influence on my research over the years. As described in this and other chapters of this book, my work on NPLs, nuclear batteries, advanced fusion fuels, alternate fusion confinement concepts, and radiation detectors have all been driven from this early insight.*

Going to an advanced fuel would not only enable direct conversion but also alleviate the problems of neutron damage to the reactor wall and the need for tritium breeding, safely maintaining tritium inventories and neutron induced radioactivity in structural materials. Also, due to the use of neutron-lithium reactions to breed tritium in the D–T approach, lithium (along with deuterium) becomes the fusion "fuel" for such reactors. Lithium, however, is not an unlimited resource, so the old ideal of fusion from an "infinite" supply of deuterium extracted from water no longer applies. In contrast p–$^{11}$B uses hydrogen and boron (from borax) which effectively offer "infinite" fuel supplies. D–$^{3}$He requires lunar $^{3}$He mining, but this still seems an important approach, especially for fusion application in space. That is another story within itself which is discussed briefly later in this chapter.

My take on this issue was that once breakeven confinement is achieved with D–T, the step up to breakeven with advanced fuels like D–$^{3}$He would not be so difficult, and would take less time to develop than envisioned. (As much time has gone on without a breakeven D–T experiment, I now favor going directly to an advanced fuel breakeven experiment as proposed by several small private companies working on fusion.) Daunting technology problems like radiation damage to the fusion vessel wall and tritium breeding would disappear with aneutronic fusion such as p–$^{11}$B. Time-wise, there appears to be a trade-off between the added physics challenge of burning these fuels versus the technology issues of D–T fueled fusion.

I ended up devoting a portion of the 1976 *Fusion Energy Conversion* book to the physics of advanced fuels and to various methods to incorporate direct energy conversion methods into such reactors. The sponsors at ERDA were concerned about this unorthodox view, and might have stopped the book project except for the enlightened support by one of their staff, Bill Gough. He personally pushed for publication. Later, Bill served as the contract monitor for an ERDA grant that I held to study advanced fuel physics. Interestingly, Bill and another far-sighted colleague at ERDA, Ben Eastlund, published an ERDA report and an article in *Science* about the use of a fusion reactor to "close the materials cycle." Their vision was to "dump" materials into the fusion exhaust plasma which would vaporize everything down to elemental constituents. The elements could be separated by a large-scale mass analyzer and recycled to a plant for use in various manufactured products. This concept received good comments, but was largely ignored by most fusion scientists as being too far off to seriously pursue. Both Ben Eastland and Bill Gough eventually left ERDA.

Ben founded a very successful company that used plasmas for coating beer cans during their manufacture and also developed some award-winning plasma lights for commercial use. Bill ended up in charge of the fusion group at the Electric Power Research Institute (EPRI), the U.S. utility research arm. There, he received advice from electric power plant executives and engineers. They told him that fusion in its present form, using gigantic D–T Tokamaks with many technological hurdles, was not of interest. Indeed, these advisors favored (largely for potential power plant attractiveness) a smaller, environmentally friendly advanced fuel plant. Thus, Bill developed several projects to study advanced fuels. I held one of the contracts. DOE, however, had a much larger research budget and generally viewed the EPRI work as unwanted and a thorn in their side. Meanwhile, EPRI was fighting many more near-term battles due to difficulties in existing fission power plants. Thus, fusion once again seemed too far off, and the EPRI fusion program gradually shrank in size. Bill left to become a contract monitor at the Stanford Linear Accelerator Center (SLAC). I lost track of him after that until around 2007 when we again began correspondence about fusion. Ben Eastlund had retired in St. Louis, MO, at that time, and he opened up conversations with Bill and then with me. Ben now had time to revive the old fusion torch concept for materials recycling and wanted to collaborate once again with Bill. He suggested involving me, so Ben then called me to discuss his plan for again studying the fusion torch concept. Ben had some thoughts about more near-term versions that could work with MHD using exhaust gases from coal fired power plants. He wondered if I would like to join them in this work. I then had several phone conversations with Bill Gough, who lives near Palo Alto, CA. I next had an appointment to talk via phone with Ben. Unfortunately, he suddenly passed away the morning our phone discussion was scheduled. I ended up with an unexpected brief conversation with his grief-stricken wife who broke the news to me. This tragic event halted the proposed collaboration, but eventually Bill Gough and I got together and are currently working on related concepts for use of a plasma jet to process municipal wastes. This concept, while much needed, is not nearly as aggressive as the old concept of closing the materials cycle, but it is easier to achieve and, being more near-term, has some funding possibilities.

Regarding the ERDA book projects, I must note another influence on my work that came from the wonderful editors at the American Nuclear Society (ANS). Norm Jacobson "held my hand" as I struggled with the first nuclear energy conversion book. This book brought in many original works and calculations that I did especially for that publication. Combining this new work with the prose needed to make things understandable, plus bringing in appropriate background work of others, was demanding. This was especially tedious in those days without word processing programs and the ability to easily change documents. My typed work had many hand-inserted marks and changes. (Typing and detail are not my strong points!) Copies to Norm came from carbon paper.

Norm kept me calm and helped smooth out sections of the book. John Graham replaced Norm on the second book. In addition to being an editor, John was the ANS representative in Washington, DC. He had gained much political savvy in this role. Thus, as the concerns about the proper coverage for the book came up, John was able to help with ERDA negotiations. He also continued, as Norm had, to help the process despite my somewhat temperamental nature when things got tough. Subsequently, Norm, John, and I became close personal friends who convened at various ANS functions, as well as several Illinois–Wisconsin football games. (Norm was a loyal alumnus from Wisconsin and a long-time season football ticket holder there.) Norm is now deceased, and John Graham retired and moved back to Gainesville, FL.

As I was writing the fusion book, I became frustrated by the lack of fusion cross section data for fusion fuels other than D–T and D–D. I wanted to include some calculations in the book for advanced fuels there was some data available but not in a convenient form for computations. For such work, so called "reactivities" or $<\sigma v>$ are used. The reactivity represents an "average" over the product of the fusion cross sections $\sigma$, and the ion velocity v. This requires a triple integration over the assumed plasma energy distribution. Traditionally because most confinement systems like Tokamaks operate in a near equilibrium condition, a Maxwellian plasma distribution function is assumed in the integrations. However, some alternate fusion concepts like the IEC operate in a non-equilibrium condition better represented by a beam–Maxwellian distribution. Averages over such a distribution were not generally available. I enlisted the aid of two graduate students, Harry Towner and Nenad Ivich, to help with developing the desired cross sections. They were very good with computer computations and were enthusiastic about fusion. After doing much work for calculations in the book, we decided to extend this cross section work farther and publish it as a fusion cross section "barn book." (The title is in reference to the famous barn book of fission cross sections published years earlier by scientists at the Brookhaven National Laboratory.) Fermi had originally given the term *barn* to these cross sections, defining it as $10^{24}$ $cm^2$. The neutron fission cross section approached that

*The cover for the Fusion Cross Section "barn" book proudly shows the sun shining over the barn.*

value and Enrico Fermi is said to have quipped that it was so large that neutron induced fission was akin to hitting a barn door with a baseball. Our *Fusion Cross Section and Reactivities* book had an image of a barn (complete with a door) on the cover, like the original BNL barn book. To depict fusion we added a bright sun over the barn. Our barn book was published as a report from the University of Illinois Fusion Studies Laboratory. By exchange agreement these reports were sent to all of the major fusion labs worldwide. We ran off an extra one hundred copies to handle individual requests. As it turned out, this was one of my more popular publications. I did not view it as a research publication, but the fact that the contents were so convenient and useful to researchers in the field made it extremely popular. We have printed almost 900 additional copies over the years and a few reprint requests come in each year. I certainly had not anticipated this, but I feel good about providing this service to my colleagues. In addition, both of the students who helped me ended up with good positions at fusion labs.

After completing the second book, I again had more time to focus on my research. As I returned to advanced fuel work, I reasoned that these demanding fuels stood a chance to succeed in high beta (high plasma to magnetic field pressure ratio) confinement systems. Such systems include Reversed Field Pinches (RFPs), Field Reversed Configurations (FRCs), Spheromaks, etc. Further, Inertial Confinement Fusion (ICF) and Inertial Electrostatic Confinement (IEC) would completely eliminate the confining magnetic $\beta$ field. I did a series of studies on a variety of these systems, and in recent years, I have tended to focus on ICF and IEC, two other confinement schemes of interest — again, no magnetic field. As time passed, many other fusion researchers became interested in advanced fuel fusion, and DOE has supported some work on this topic. Several privately funded companies, EMC$^2$ and Tri-Alpha have put the achievement of p–$^{11}$B fusion as their stated objective. However, the vast majority of the international fusion community remains "glued to" D–T and Tokamaks (or D–T and ICF). It will be interesting to see how all of this plays out in coming years. Hopefully, fusion will gain a prominent place in the search for a future energy solution. However at present, due to the perceived long development time required, fusion is not considered relevant when the energy crisis is discussed.

My work on advanced fusion fuels was reinforced when, by chance, I accepted an invitation from Manfred Heindler (then Professor of Physics at the University of Graz in Austria) to speak on the subject at the first International Meeting on Emerging Nuclear Energy Systems, held at the University of Graz. Liz accompanied me, and we greatly enjoyed Graz with its friendly and welcoming people, beautiful scenery, museums, and historic buildings. Winnie Kernbichler, a young Assistant Professor at Graz, helped Manfred with the meeting organization. One important result of the meeting was its effect on subsequent research at Graz. Manfred and Winnie both became enthralled with the possibility of advanced fusion fuels. Also, they saw it a way to create a niche for fusion research at the University of Graz which, up until then, had little research or courses in the area.

I was not the only one who spoke about advanced fusion fuels at the meeting. Bogdan Maglich was studying a novel colliding beam device (the "Migma" fusion concept) which claimed to be near breakeven with D-$^{3}$He fusion. He gave a typically (for him) flamboyant talk. Bogdan was funded privately and had a small laboratory in Princeton, NJ, where he was a thorn in the side of Princeton Plasma Physics Laboratory scientists and DOE officials in general. He claimed — loudly, including lobbying congress — that his approach could succeed with only a fraction of the money being spent on, as Bodgan called them, "non-promising" DOE concepts. In this spirit, he boldly announced at the meeting that he was near to securing funding to build a fusion "Migma research lab" in Graz. He predicted that their research would rapidly lead to a fusion power plant. Bogdan was born in Austria, so doing this there made sense to him. A reporter at the conference then released a second page headline in the local paper, stating "Scientist Plans to Build Nuclear Power Plant in Graz". The typical person on the street did not know the difference between fission and fusion (not specified in the article), but there was at that time strong opposition to nuclear fission power plants in Austria. Thus, the next day, meeting attendees had to fight their way through picket lines, where protesters held signs saying, "No Nuclear Power Plants Here", "Stop the Nuclear Power Plant", and so on. That was the last I heard about the proposed Migma research lab in Graz. Sentiment about nuclear power has changed over the years, but I doubt that many yet distinguish between fusion and fission.

Bogdan Maglich and I remain "friendly" colleagues to this day. But, that friendship has been fragile. In the 1980s, Art Gunther, Chief Scientist at the Kirkland Air Force Base in Albuquerque, NM, recommended that I chair a special committee formed under the Air Force Research Advisory Board to review Bogdan's research, which was then funded by the Air Force. His aggressive claims were exciting but raised much controversy. After some arm twisting, I reluctantly accepted this task. Midway through the review Bogdan became incensed over some questions asked by the committee and phoned me, threatening a lawsuit. As it turns out, he was not intending to go to court, but this was his way of saying he violently disagreed with the committee's preliminary conclusions (provided to him for comment). When all was said and done, after much deliberation and debate, the final committee report was neutral (encompassing a wide divergence of opinion) about his work. In my opinion, sometimes it takes radical people like Bogdan to move the system forward so that people can focus on the true mission of finding a Black Swan, versus aimlessly wandering in the forest.

Speaking of getting lost in the forest, that happened to me when I wrote the book *Direct Conversion of Nuclear Energy* in 1967 under a contract from ERDA and published by the American Nuclear Society (ANS). At the time ERDA officials were mainly thinking about nuclear pacemakers, which were widely used then due to the long lifetime offered (versus chemical batteries). However, as chemical batteries improved and offered much lower costs, nuclear-powered pacemakers soon faded from view. ERDA wanted to document the basics of direct conversion involved in pacemakers. However, this concept excited me, especially because the nuclear-pumped laser was also a nuclear source giving direct conversion to optical energy. Thus I realized the concept was much broader than radioactive pacemakers. As a result, I enthusiastically did considerable research to develop both a basic process description and include a number of applications. The latter included things like the Gamma Electric cell; various types of nuclear batteries involving betas, alpha particles, neutron reactions; the fission electric cell; nuclear-pumped lasers; and radiation induced plasma cells. But my personal focus at the time was on nuclear power production. Thus I viewed my chapters on the fission electric cell (converts fission fragment energy directly to electricity based on fuel element designs with thin coatings of uranium so that the alphas could escape the surfaces at high energy) as the key contributions of the book. However, I was intrigued with and did a "first" in-depth treatment of the physics of the interaction of high energy radiation (or energetic particles from neutron induced reactions) with matter for direct conversion to electricity.

After the book came out, isotope-powered pacemakers were rapidly going out of use. Because this was a major use for direct conversion at the time, I thought the other parts of the book would only excite a few futuristic scientists. Those scientists might be interested in advanced fission reactors using the fission-electric cell approach in reactors or delivering optical output via a nuclear-pumped laser. Thus it was a real surprise when a year later I found out this book was the leading seller in the ANS book series, a position it retained for some time. What had I overlooked? Well, it turns out the basics discussed contain a wealth of information for people interested in nuclear radiation detectors — such detectors essentially involve conversion of radiation energy through interactions with matter to produce charged particles that provide the electrical output signal. The amplitude of the output signal depends on the radiation intensity, providing the basis for the detection instrumentation. I knew this, but I had let it slip from my view during book preparation. Fortunately, I did not overlook the basic processes. I partially "recovered" from this oversight when a few months later a senior scientist from Reuter Stokes Co. (a major manufacturer of nuclear detectors for use in fission reactor instrumentation in Cleveland, OH) called and asked if I would consult with them on the design of a new detector series they were considering. It would involve some of the direct conversion concepts described in my book. The detector would be very small and avoid current saturation even when inserted in the very high neutron fluxes in the center of a power reactor core. Another version would be designed to use in pulsed power reactors such as the TRIGA. The detectors could discriminate between neutrons and gamma rays. I enthusiastically agreed to work with them on this challenging application of some of the basic principles provided in my book. The series of detectors was termed "Semi-Rad," a trademark of Reuter Stokes. I had missed the train (maybe even the Black Swan) by not stressing that the direct conversion I was talking about in the book was so fundamental to instrumentation. My tunnel vision on power production had blinded me. Perhaps I should have titled the book *Radiation Energy Conversion for Nuclear Detectors* — then interest and sales might have been even an order of magnitude more. (Sales numbers were important to me only in terms of interest in the book by the community. ERDA paid me a flat fee for the book — equivalent to a fraction of a cent per hour for the vast time I put into it. That didn't bother me, because one doesn't write such a book for money!)

Some years later, I discovered yet another interest in this book that stemmed from the Gamma Electric Cell analysis presented in it. I was asked to visit LANL for a classified discussion (as noted elsewhere, I held a DOE security clearance for many years at LLNL for ICF physics, and it enabled my entrance to classified areas at LANL also). The discussion involved some equations I had derived in the direct conversion book and work I had done with Henry Sampson. I was impressed to see my book on the desks of many scientists in the group of 20 or 30 people I was visiting. My clearance restrictions prevented me from getting a clear view of the purposes of this interest, but it clearly involved conversion of energy from a nuclear explosion. I gathered some underground tests (OK in those days) being planned to study the concept. That visit seemed to clear up the LANL group's questions involving me, and I lost track of efforts thereafter. But, I surmise the concept is now an option that could provide an ultra high voltage power source during a nuclear explosion. The exact use of such a voltage source is not clear, however.

My work on advanced fuel fusion is closely tied to direct energy conversion because one reason to go in that direction is to obtain sufficient energy in charged fusion products to make direct conversion attractive. The University of Graz meeting where I spoke about this tie resulted in a close association with Manfred and Winnie that lasted for many years. I fondly remember several non-technical events during the years of our association with Manfred and Winnie. One summer, while I was at LLNL, Manfred and his wife Claudia "house-sat" our home while Manfred did research at the University. They fell in love with our little West Highland White dog who they fondly nicknamed "White Flash." Manfred, as did many Europeans, was enamored with the huge American cars (sedans and station wagons of that period). He also wanted to see America. So when I returned from LLNL, Manfred promptly purchased a used Lincoln sedan, and he Claudia set out on a great circular road trip through the West. Recounts of the adventure include a vivid description of a dozen delays due to problems (breakdowns) of the car, including a boiling radiator, multiple flat tires, etc. Manfred had the foresight to buy a tool box and gloves which found good use! They adored the national parks, such as the Grand Canyon, Yosemite, Yellowstone, and the Badlands.

Several years later, Winnie invited me to an IAEA winter school on fusion that he ran at a famous ski resort in the high Alps (see the earlier photo in Chapter 7). I enjoyed the technical part, interacting with various old friends and also European scientists I had not previously met. And, I think I set some sort of ski record for the most falls per run! That is what happens when someone from the Illinois prairie who has mainly done cross-country skiing attempts downhill skiing in the Alps!

Subsequently, both Manfred and Winnie spent sabbaticals with me in Illinois, and Liz and I visited them again to enjoy time in Graz, continuing to explore that city and the surrounding countryside. Austrians are fun-loving people, and Manfred and Winnie set a good example of this during our visits. We had some fun times in addition to doing joint research. Manfred later became the Minister of Energy for Austria, moving from Graz to Vienna. Thrown into that position, he began fighting the day-to-day problems, as many government officials worldwide must do. Fusion, as a longer range program, faded from his radar. He told me he planned to eventually leave government and return to fusion research. However cancer dealt a fatal blow before he could do that. Winnie progressed to full Professor and covers a wide range of physics, including fusion.

Another young staff member at Graz, Hans Reidler, also came to Illinois for a year to work on advanced fuel fusion. A niche he and others at Graz concentrated on was to fill in gaps in cross sections for several of the advanced fuels by extrapolations of existing data. Their work was intended to update and extend the fusion cross section book I had previously published through our Fusion Studies Lab report series. Their results were eventually published in a thick IAEA report on the subject. Hans left the University to work for a private company, and I had lost track of him, until one day he called to ask a favor. He had become emotionally involved with the horrible killing and destruction going on in neighboring Bosnia. Hans had then tried to help people with shelter and first aid, risking his life by going into war-torn areas. But he and some like-minded friends had run out of supplies and money. Always a blue sky thinker, he came up with the plan to convince Tina Turner to give a concert to raise money. He asked if I could contact her about this. Unfortunately, I confessed to Hans that I didn't know who Tina Turner was (I don't follow many popular performers), and would not have the slightest idea of how to contact her. We talked then about other charitable aid organizations he might approach. Again, I lost track of him, but later learned that Hans managed to get more supplies and continued to "do good" until he had to return to his job (he had taken a 10 month "vacation" from it to help people in Bosnia — what dedication).

As illustrated by my experience in Graz, advanced fuels are an extremely attractive and challenging approach to fusion that can fascinate people interested in fusion. Thus over the years a number of students working with me were drawn to this area. Typically they then moved into some other "enabling" aspect, such as studying alternate confinement methods which would be best suited for burning such fuels (Chapter 8 on alternate confinement fusion Chapter 11 on IEC fusion discuss this). Others, like Phil Stroud, Terry Chu, and Harry Towner, worked directly on advanced fuel plasma physics and associated fusion cross sections.

In addition to electrical power, the use of fusion in space is very intriguing. Fusion-powered propulsion is generally viewed in the space program as a leading candidate for deep space missions. The same attributes that make advanced fuel fusion so attractive for use in electrical power plants carry over to use for a space rocket. In that case, most of the fusion power goes into energetic charged particles (reaction products) that can be exhausted out through the rocket nozzle to achieve thrust. I have been one of the leaders in exploring such concepts and have written a variety of papers on the subject, as discussed later in Chapter 14.

Another important influence on my advanced fuel and energy conversion work came from my long time association with Hiromu Momota. I first met Hiromu after a professional society meeting in Japan, shortly after the publication of the *Fusion Energy Conversion* book. He asked me if we could meet privately to discuss some aspects of the book and told me that he had plans for a burning plasma experiment. Hiromu had read the book closely and was slowly becoming convinced of the need to move to advanced fuel fusion, especially D–$^3$He. He said that the Advanced Space Projects Agency of Japan had become very interested in the discovery of plentiful $^3$He on the moon and its potential use in fusion reactors. Apparently, this agency had formed a group to study the situation and make recommendations to the government. Hiromu was a senior Professor of Physics at Nagoya University where he worked in their widely known plasma laboratory. He had participated in the Advanced Projects review and was now planning to create a project to perform research on the topic. He had already organized a group of outstanding staff from other universities to join him in the study. Subsequently he held several Japan-U.S. "specialist" meetings on D–$^3$He, which several American colleagues and I attended. Hiromu's multi-university group soon received support to undertake a quite detailed design study of a D-$^3$He fueled FRC with a Traveling Wave Direct Energy Convertor (TWDEC). This project retained the original study group and now pulled in participants from several major Japanese companies in Japan. Several "start-up" meetings were held in Japan. I was invited to attend along with several other U.S. participants, especially FRC specialists from LANL and the University of Washington. The final report on this project is still one of the most important fusion reactor studies to date. It provides the vision of a very attractive fusion power unit named "Artemis" which is cost effective, efficient, and environmentally compatible.

Hiromu later retired from Nagoya University and came to Illinois to work with me. (Japan's universities have fixed retirement ages and do not permit continuation beyond that time unless one goes to a "private" university or college or industry. Thus, options are very restrictive, but hopefully will change in the future to take advantage of this key "brain power." Anyway, Hiromu felt that coming to Illinois was the best way to keep heavily involved with students and with fusion research.) The original thought was that we could redo the energy conversion book and co-author an updated version. However, we became so involved with new joint research projects that we never had time for the book revisions. This association lasted over six years, until Hiromu and his wife, Michiko, returned to Japan in 2007. Our work had resulted in several DOE and NASA contracts, and many papers on advanced fuel FRCs and also IECs. In addition to the research, Hiromu was instrumental in helping me advise several PhD students. Hiromu and I still remain close friends, but via e-mail. I remain most grateful and blessed for our time together.

During the time of my work on advanced fuel fusion and its space applications, I was appointed to the Air Force Research Advisory Board and served on it for about 10 years. In a sense, this appointment grew out of the panel I chaired for the Air Force review of Maglich's Migma project (described earlier in this chapter). This board was administrated through the National Academy of Sciences. Meetings involved discussion and planning with many high-up scientists and generals in the Air Force. A memorable occasion for me was a meeting at a building adjacent to the State Department in Washington, DC. This was during the time of Desert Storm, the "first" war with Iraq, following their invasion of Kuwait. We adjourned the formal meeting for a dinner. I was seated near an Air Force Brigadier General, the senior military person attending. Halfway through the meal, an Air Force Lieutenant broke in and softly told the General that "CNN says we have bombed Bagdad" (strange source for information considering Air Force intelligence). The General made a hasty retreat to return to his office and suggested we end the dinner quickly. When I went to exit the hall, I found the route blocked by security guards. Outside the local streets were blocked off due to fears of possible reprisals against the State Department or others in that location. It took me two hours to get through all of the check points and return to my hotel. Fortunately, I was leaving, not trying to enter!

There were a number of other interesting events in association with my service on the Air Force Research Board, but space does not permit me going into all of them. Because I am working on LENR, a form of cold fusion, I cannot resist telling about one event that concerns cold fusion. The research board was periodically briefed on progress by directors of the Air Force, Army, and Navy research directorates. During the year after the Pons-Fleishmann announcement (discussed later in Chapter 12 on LENR), the director of the Office of Naval Research (ONR) was asked during his briefing what the major achievement was for the year. He responded, "I stopped all of that cold fusion nonsense." I was caught off-guard, not even thinking about the Navy doing cold fusion research. I then said "that seems like a 'negative' achievement — what would you have for a positive one?" Anyway, even today, administrators in many government agencies feel cold fusion to be "nonsense" without having truly investigating it. I would add that later I found out that scientists at the Naval Research Lab in Washington, DC, were in fact doing (and continue) cold fusion research. It is not widely publicized so as not to raise undue concerns by administrators. Apparently the Director of the ONR did not know that when he briefed the Advisory Board! David Nagel, then Head of the Solid-State Division at NRL was the main supporter of this work and he got it started there with internal funds. He has since retired from NRL and is a Professor at George Washington University in DC. Indeed Dave was a co-organizer of a meeting in the series of international conferences on cold fusion (ICCF series) held in DC in 2008.

Recently John Scott and others at NASA Johnson Space Flight Center in Houston have become very interested in aneutronic fusion for space propulsion and in the Traveling Wave Direct Energy Convertor (TWDEC) for energy conversion. He is planning a TWDEC experiment to extend Japanese work to higher density and voltage operation. I have obtained a small NASA grant to assist in this effort. The hope is that it will build interest in NASA which may eventually move to future aneutronic fusion propulsion studies. Much depends, however, on how future congress members handle NASA funding now that the shuttle program has ended, and more emphasis has been placed on involving private companies in the space program.

## Further Reading

G. H. Miley, *Radiation Energy Conversion*, ANS, LaGrange IL, 1967.

G. H. Miley, *Fusion Energy Conversion*, ANS, LaGrange IL, 1976.

G. H. Miley, "A Review of Direct Conversion of Nuclear Energy," *Fusion Technology*, Vol. 20, p. 979 (1991).

G. Miley, H. Momota and J. Nadler, "Direct Energy Conversion for IEC Fusion for Space Applications," *Proceedings 36th AIAA/ASME/SAE/ASEE Joint Propulsion Conference and Exhibit,* Marshall Space Flight Center, Huntsville, AL, Jul. (2000).

G. H. Miley, H. Towner and N. Ivich, "Fusion Cross Sections and Reactivities," COO-2218-17, U.S. Atomic Energy Commission (1974).

H. Momota, A. Ishida, Y. Kohzake, G. H. Miley, S. Ohi, M. Ohnishi, K. Sato, L. C. Steinhauer, Y. Tomita, M. Tuszewski, "Conceptual Design of the D-$^{3}$He Reactor Artemis," *Fusion Technology,* Vol. 21, p. 2307 (1992).

# Chapter 10

# Inertial Confinement Fusion (ICF)

*Upper Left: Heinz Hora, Professor of Theoretical Physics at UNSW, Australia, got me interested in ICF and has been a friend and collaborator for many years. His wife, Rose, shown here, was always a stabilizing influence in his hectic life, which involved trips to many professional society meetings and invited lectures around the world. He set the*

*record for travel in my view. Rose passed away from cancer in 2005 after a long struggle. That left a tremendous void in Heinz's life, but hasn't stopped his research productivity or traveling. Upper Right: I was honored to have Edward Teller personally present the Teller Medal to me during a fusion meeting in Japan. As described in the text, he made the trip just after recovering from some medical problems. Bottom: I'm in the front row as organizer of the periodic international workshops on laser-particle interactions held in the summers at the Naval Post Graduate School in Monterey, CA. These conferences attracted many outstanding scientists in the field. I co-authored, with Heinz Hora, over a dozen proceedings from these workshops published by Plenum Press and later by AIP. These workshops eventually grew into the current Inertial Fusion Science and Application (IFSA) series, which rotate between the United States, France, and Japan. The size of the present meetings is an order of magnitude larger than the old Monterey series. I am proud to have been involved at their start. I am on the IFSA advisory board, but no longer play an active role in the conference organization.*

*Left: Professor Sadao Nakai, then head of Osaka University Laser Fusion Laboratory, and I are shown discussing some ICF target physics during a break at an ICF meeting. We became quite close friends and he later invited me to participate in several small group meetings at the Osaka Laboratory to discuss future plans for laboratory development. Right: I attended a number of laser fusion meetings in Japan. I am shown here making a presentation in 1992 at a Kyoto meeting on laser matter interactions.*

My interest in Inertial Confinement Fusion (ICF) was a natural outgrowth from my early work on lasers. ICF was originally termed "laser fusion" until it was realized that other pulsed power sources like heavy ion beams could also be used. The use of intense lasers to focus on and compress a small target of fusion fuel was the first vision of the ICF program and was reported by Teller, Nuckolls, Wood, and colleagues for the first time in an open (unclassified) meeting at an IAEA conference in Toronto, Canada in 1970. In later years, interest also grew in other intense pulse power sources (such as accelerators, z-pinches, etc.) could be used instead of lasers and might offer an efficiency advantage. My interest in ICF was further intensified in the 1970s when Keeve (Kip) Siegel founded KMS Fusion in

Ann Arbor, MI. Kip had been a Professor of Electrical Engineering at the University of Michigan when I was in graduate school there. He formed KMS Electronics, which became a major private company in the U.S. doing business with military contracts. However, for some reason, he became extremely enthusiastic about ICF and sold off most of his electronics business to gain money for KMS Fusion. His plans were somewhat secretive, but I gradually learned of them. It turns out that I personally knew many of the people working on the project. Henry Gomberg, former Head of the Nuclear Engineering Department at the University of Michigan, was the senior scientist on the project. He had previously been on my thesis committee. Gomberg proposed some schemes for using the alpha particle energy created in D–T fusion reactions to promote chemical reactions. Indeed, the concept was to use the KMS Fusion project to develop a hybrid fusion plant that would both produce electricity and manufacture a synthetic methane-like fuel. Wayne Meinke was placed in charge of developing methods for filling micro-balloon targets with D–T for use in the laser implosions. He was a consultant to KMS and a Professor of Radiochemistry at the University of Michigan. Indeed, Meinke had instructed my radiochemistry lab class at Michigan. The students in the class joked about it doubling as a gym class. We often had to make mad dashes to get short-lived isotope samples generated in the Ford Nuclear Reactor to the counting room. Meinke came up with a very innovative target concept. Glass micro-balloons were heated up in a gaseous deuterium/tritium atmosphere so that the hydrogen isotopes diffused into the center of the gaseous balloon. When cooled down the glass walls had a very low diffusivity for these gases, so they were trapped in the interior, forming an ideal target. This basic concept was later used throughout the ICF target industry, but was not well known at that time. The trick was to find or make extremely symmetrical sub-millimeter sized glass micro-balloons. KMS scientists found that they could buy beakers full of these sub-millimeter sized balloons from glass companies who formed them by a liquid drop process for use in reflecting paints for highway signs. The only problem was that the glass shells were all too often non-spherical or possessed non-uniform wall thicknesses. Initially, only one out of a hundred of the micro-balloons purchased fit these demanding requirements for extreme symmetry. The tedious selection process was done by hiring a large number of students who sorted micro-balloons through visual inspections with microscopes. Today, the manufacturing processes have been modified to specifically make balloons with close tolerances so that the sorting is much simplified. However, that technology was not available in the days of KMS's work.

Another very innovative concept that Kip Siegel and his colleagues came up with was the use of a clam-shaped mirror to provide reasonably symmetric $4\pi$ illumination of the target, with only a couple of laser beams entering the chamber. At that time, other labs were still struggling to form and focus numerous separate beams on the target to gain symmetrical irradiations. However, that work was very slow due to the simultaneous demand for obtaining extremely precise timing of the beam arrival along with good focus on the target for all the beams. Finally, to implement these innovations in a research effort, KMS contracted for an advanced glass laser with a pulsed power system from the French General Electric Company. The company representative assigned to this project was Jean Guyot — the same Jean who had done his PhD on NPLs with me! As a result of my association with Jean, Wayne Meinke, and also Henry Gomberg, the KMS staff, and Jean himself fondly kidded me by calling me the "Grandfather" of the project. All this technology fell into place rapidly, and KMS got some excellent initial results with their innovative setup. Thus, they became very optimistic that they could achieve a breakthrough with fusion. That year, Kip Siegel sent out a Christmas card which carried an enthusiastic message that KMS Fusion would supply synthetic fuel gas ("syn fuel") for use in an advanced gas-fired power plant within ten years! Kip seemed to be overly optimistic, but he certainly got things going with his enthusiasm.

To make a long story short, KMS Fusion no longer exists due to a variety of unfortunate circumstances. First, Kip Siegel, the master "salesman" for the project suffered a fatal heart attack during an emotional testimony to congress about the project. His amazing money raising abilities were then lost to KMS. Meanwhile, the National Laboratories were slowly making progress in ICF and, with superior financial support, were bypassing KMS accomplishments. Simultaneously, a lawsuit developed claiming that one of the KMS consultants, Keith Brueckner, had violated classification restrictions in some of the technology employed at KMS. It is not completely clear, due to confidentiality at KMS, exactly how the syn fuel production plant was envisioned to proceed. However, without Keith to explain the program, worries developed that the time scale for success was nowhere near the optimistic projection of Keith's Christmas card, and investors were impatient. To gain income, KMS Fusion turned to manufacturing glass micro-balloon targets for the ICF national programs. Indeed, they were the sole supplier for several years. Then, some political infighting developed and the contract was abruptly stripped from KMS Fusion and given to others advised by, ironically, one of my former colleagues and friends from the University of Illinois, Chuck Hendricks. Chuck was a Professor in Electrical Engineering at the University of Illinois and a leading researcher in forming electrically charged droplets of liquids, including glasses, printer inks, etc., and guiding them electrostatically for applications like printing. Subsequently, Chuck turned to fusion applications of this technology. He obtained a DOE project which culminated in injecting, for the first time, fuel pellets into the ORMAC Tokamak fusion experiment at the Oak Ridge National Laboratory (ORNL). He next turned to ICF targets and was subsequently lured away from Illinois by LLNL to head the ICF target group there. Later, Chuck was a founding member of a private company specializing in ICF targets. He advised General Atomics (GA) located in San Diego, CA, on how to get into the area. Shortly thereafter GA took the DOE target contract away from KMS Fusion. The loss of that contract ended KMS's work on fusion. KMS Electronics continues to work in the old fusion facility, but are now doing various electronics projects along the lines of the original company before Siegel took it into fusion.

Sometimes events seem to go in surprising circles. In this instance, although I was an "outsider," I knew and felt close to all of the key persons involved in ICF target research. I was looking in on my associates and friends as they sought their own personal Black Swans. The development of a successful ICF target "fuel factory" could truly be a Black Swan in terms of achieving the most unbelievable combination of technology and economics. The owner of the fuel target factory (i.e., the fuel for ICF reactors) may well have the most profitable part of the ICF power plant business. It's akin to supplying coal, oil, or natural gas to current power plants. No wonder this is a dream area in the minds of many people and companies. But, the winner of this Black Swan remains up for grabs. Again, ironically, I got into the chase personally in recent years by proposing a non-cryogenic target based on our recent deuterium-cluster technology discussed in Chapter 12. This approach, while intriguing, seems a "long shot" due to many uncertainties. But, as described in Chapter 12, we have gone so far as to perform some experiments based on a variant of this target concept using the TRIDENT Petawatt laser facility in LANL. In this case, however, our goal was to use a "convertor foil" to create an intense deuterium beam during the Petawatt laser interaction with the deuterium clusters in the foil. This beam could be used for fast ignition of the ICF target as discussed in Chapter 12. Thus the concept is different from the original use of clusters for the target itself. However, one never knows when a Black Swan event may happen so we must keep forging ahead.

An early event that propelled me further into ICF studies was a workshop on ICF, chaired by Professors Heinz Hora (University of New South Wales) and Helmut Schwartz (RPI), held at the RPI Campus in Hartford, CT in 1972. At the time, little was known about laser plasma interactions and ICF was in its infancy. The two of them had formed a close scientific association after Heinz offered a theoretical explanation for the controversial "Schwartz effect" — but that's another story. I had many interchanges with Heinz Hora on ICF and a close collaboration that has lasted throughout the years. Meanwhile, I also met Nicholas Basov, who received a Nobel Prize for his pioneering work on neutron production in ICF experiments at the Lebedev Laboratory in Moscow.

Another person who caught my attention at this time was Moshe Lubin at the University of Rochester. He was studying laser target methods to create the initial plasma inside a mirror confinement system. This intensified my interest and, in 1969, I decided to contact Moshe for a discussion about possible collaboration. He discouraged me, saying that his experiment seemed to be a dead end. I was extremely surprised when it was announced a few years later that the DOE awarded him a large grant to create an ICF facility at the University of Rochester. It was named the Laboratory for Laser Energetics (LLE). This facility would have the largest ICF laser in any university in the United States, and was justified as a combination of a DOE research facility and a DOE-supported national user's facility. As a result, Moshe began to focus on getting an initial group of users ready to carry out advanced ICF experiments on this exciting new facility. He remembered my interest in his work and flew to Illinois personally to discuss the possibility of me becoming a user. This impressed me greatly, because normally I would write a proposal to become a user and wait for a committee to review it. My colleague, Chan Choi, and I had been studying alpha particle slowing down and possible instabilities in ICF targets from a theoretical point of view, so I proposed to do experiments for comparison with the theory. We did write a proposal to do this work, and with Moshe's endorsement it sailed through the review process. The Rochester facility laser was capable of reasonable fusion yields using shock implosion of glass micro-balloon targets. While this type of implosion is not scalable to a future power producing target, it was an expedient way to get the experimental yields of interest at that time. I proposed use of a magnetic spectrometer to measure the energy of alpha particles escaping from the fusing targets. In fact, in order to cancel out some instrument inaccuracies, I suggested the simultaneous measurement of the energy spectrum of alphas and protons created by D–D and D–T fusion reactions during implosions of deuterium rich D–T targets.

With my proposal for the experiment at LLE accepted, I began design of the magnetic spectrometer needed to measure the particle energies. Not long after my design was nearing completion I attended the opening ceremony for LLE at Rochester, where we enjoyed dinner at tables situated around the large target reaction chamber with its laser and diagnostics ports sticking out in all directions. It was at that moment that I realized my bulky magnetic spectrometer would not fit between the chamber supports! This made the whole thing impractical. Fortunately, one of my old friends from KMS Fusion, Don Slater, saved the day by offering to assist me by providing the specifications for a design for a time-of-flight spectrometer which had been developed at KMS. Don was the son of Marlo Slater, a good friend from our church in Champaign. I also know his brother Jack, who also went into science and ended up as the technical monitor for one of my Air Force laser research contracts.

The time-of-flight unit did not quite offer all of the capabilities of my original magnetic design, but it was good enough and would fit in the space allowed. Fortunately, one of the people who had worked at KMS on the original spectrometer, Aaron Bennish, had come to Illinois to study in our department. He joined my ICF group, which included two other graduate students, to work on this project. His knowledge of the spectrometer enabled us to move ahead much more rapidly than we would have otherwise. The other students, Dale Welch and Dave Harris, were also very motivated and talented. Another Assistant Professor that I hired in the fusion area, Finis Southworth, became involved in the theory along with my long term collaborator, Chan Choi. The project turned out to be exceedingly successful, giving us some of the first data on fusion product spectra emerging from an implosion. This led us to suggest it as a method of measuring the $\rho R$ of the compressed target (i.e., a measurement of the degree of compression achieved) based on the ratio of 14-MeV neutrons from D–T reactions to 2.45-MeV neutrons from D–D reactions. This is a method commonly used today. The three students received PhDs on various aspects of ICF fusion that evolved from the project. In fact the development of the $\rho R$ measurement technology was not my original goal, but more or less was still a Mute Swan. From a personal standpoint, however, I was disappointed, because within the inaccuracies of the experiment we couldn't find the fusion product instabilities that we had predicted. (With respect to my disappointment, I should note that the instabilities I am discussing could be beneficial by providing an increased heating ratio for the target.) The instabilities may have been there, but with such small amplitude that we could not measure them. I believe it is more likely that we needed to change the target conditions from those employed in these very early experiments. Such instabilities are very nonlinear and might grow rapidly under certain conditions. So, in my mind, this remains an open question.

My ICF research continued after the Rochester experiments, but another aspect of my interest in this area developed when Heinz Hora got me involved in two different external positions — as Associate Editor for the Cambridge University press journal, *Laser and Particle Beams,* and as the Co-Chair for the newly initiated semi-annual international meetings on "Laser Plasma Interactions." Later, I became Editor-in-Chief of the journal and enjoyed working on it for more than a decade. The laser interactions meeting, which was by this time traditionally held at the Naval Post Graduate School in Monterey, CA, became an important event in my life for several decades. As the American host, I assumed leadership, including all the arrangements and organization of the event. Heinz remained a driving force in planning and provided an excitement in the activities and discussions with many comments and questions. Preparations were complicated by having the conference so far from Illinois, but that was soon forgotten due to the excitement of continued interactions with leading scientists in the field from throughout the world. The meeting became noted for having first time presentations of important new results. We ran this meeting at the Naval Post Graduate School because our associate there, Professor Fred Schwirzke, could obtain conference room space and housing at low cost. Meanwhile, the setting was extremely attractive and attendees always looked forward to coming there.

During these meetings, we managed to initiate the Edward Teller Award for contributions to ICF and fusion science in general. Heinz Hora was instrumental in doing this, even arranging for the casting of the beautiful Teller metals in Australia and convincing Edward Teller to allow the award to go forward. Edward Teller wanted to play a personal role in the selection of the winner for this award, and he retained the authority to review the final candidates proposed and personally make a selection. As the second Chair of the Awards Committee (Heinz was the first), I visited Edward Teller several times and always found his knowledge, insight, and enthusiasm to be keen and inspiring. He was one of those people that one felt free to talk to, though he could be intimidating with his keen insight and questions. Edward also traditionally came to the meeting the day of the award banquet and personally participated in the awards ceremony. We usually managed to convince him to give a talk. Traditionally, I would receive a call from Edward a month before the meeting where he would get around to asking me what I thought the audience wanted him to talk about. I always tried to respond, but I quickly learned that his talk seldom touched on anything that I had suggested! In all cases, his talks were captivating. We wanted to include them in the conference proceedings, which we published after each meeting, but he didn't have a written copy to give us. Thus, we recorded the talks, and I ended up with the task of preparing a manuscript from the tape. His accent combined with the sometimes poor acoustics of the tape, made this a real challenge. Thus, I had to ad lib some. My transcripts were always sent to him for review and usually came back with huge black marks throughout, especially in places where I had ad libbed to fill in what I thought he was saying. Later, this award was transferred to the American Nuclear Society and its Fusion Energy Division to handle. This ensured it would have a permanent sponsor along with financial stability. I was very proud to be one of two recipients of those early awards. Mine happened to be presented at an international meeting in Japan. Edward Teller was quite elderly at the time, but was still determined to come to Japan, attend the meeting, and present the award. This was a real challenge to him in several ways. First, it is a long flight for someone in frail health. Second, and even more of an issue, was that he had not previously visited Japan. As the "father" of the hydrogen bomb, some of his advisors feared he might receive a bad reception from some more radical Japanese. Characteristically, he ignored both issues and came ahead to Japan. Fortunately, he always seemed to gain strength when he needed it. Plus, the meeting was very pleasant for him. Rather than being resentful, the Japanese warmly welcomed him. Having him personally hand me the award was an honor which I will always remember.

The Monterey meeting series, which Heinz Hora and I co-chaired were eventually replaced by the new IFSA (Inertial Fusion Science and Applications) series, which is rotated bi-annually between the U.S., France, and Japan. Our Monterey Meeting had top people, but the attendance during those days was between 150 and 200. The Sixth IFSA meeting was held in San Francisco, CA, in 2009 with the National Ignition Facility (NIF) group at the Lawrence Livermore Laboratory serving as hosts. Over 650 people attended this meeting. Of course, some of the growth is due to the fact that the field has greatly expanded, as witnessed by the new large laser facilities in all three of the hosting countries. Indeed, it was at this meeting that I learned of the term "Black Swan" from John Nuckolls as I explained in the introduction to this book. Another memory from this IFSA meeting is that I could not help but feel the warmth of appreciation when Heinz Hora and I were asked to stand during the meeting banquet to be recognized as the early initiators of the meeting series.

One interesting consequence of my ICF target physics work developed as the fusion emphasis at LLNL shifted from the dying mirror program to laser fusion (i.e., ICF). Eventually, LLNL gained DOD support to construct the largest laser fusion facility in the world — the National Ignition Facility (NIF), designed to deliver an amazing 2 MJ of blue laser light on target. DOD wanted this facility for classified radiation physics studies — emissions from the imploded target can "mimic" conditions in a hydrogen bomb test. In an effort to move ahead on NIF construction, the Environmental Division at Argonne National Laboratory (ANL) received a contract from the Nuclear Regulatory Commission (NRC) for the NIF nuclear hazards site evaluation. The ANL staff included experts on radioactive hazards, but they knew little about laser fusion. They were aware of my ICF work and because I was located nearby, making communications easy, they subcontracted me to do the laser chamber and target part of the analysis. They assured me that this was routine, but it still needed a detailed analysis, especially the neutron activation of materials during target implosions and the tritium handling, the tritium being used for D–T targets. I gladly accepted the responsibility, having talked to my friend Mike Campbell, then Head of NIF. He had agreed to provide full support of my effort. I also enlisted the aid of one of my bright PhD students, Maria Petra, who worked on lasers (NPLs). She had an ideal background for the study, having taken our nuclear engineering courses as well as my ICF fusion course.

We worked hard on the hazards analysis, with emphasis on possible issues related to tritium handling, used in D–T target shots to demonstrate ignition. We also studied neutron induced radioactivity in structural materials as well as in target materials distributed throughout the chamber after an implosion. None of the radiation involved seemed serious if proper precautions were taken, and LLNL staff had done an excellent job in the design to cover these issues. Maria and I were relieved when we finally finished the report and turned it over to ANL to forward to the NRC. I thought the job was over then. However I had not anticipated the uproar that "concerned citizens" groups around Livermore would cause. Their statements ranged from fears that the tritium danger would severely damage their property and home values on to arguments that this weapons facility should be banned as part of the weapons test ban. They went to court and to congress with their arguments. The next thing I knew I was spending much time testifying at heated hearings — not something that I relished doing. Finally, these public hearings ended and the site analysis was approved, so NIF could move on to the construction phase. As I said in the introduction to this book, NIF was supposed to prove ignition in early 2012 using D–T targets, but has encountered some problems that will require more time to overcome. Some local citizens still complain on call-in radio and TV shows. I recently reminded Ed Moses, the current Director of NIF, that I did the NIF site nuclear hazards analysis, but he was not aware of that (done before he came to NIF). He still thinks of me mainly as an academic researcher, not a practical worker who does site safety analyses. Yet, as discussed in Chapter 17, I have done somewhat related work at the Clinton fission power reactor and on the State of Illinois Radiation Protection Advisory committee.

Meanwhile, my research in ICF has taken several directions in recent years. Heinz Hora, with my help, pioneered the concept of "block ignition" as an alternate approach to the currently popular fast ignition of high gain ICF targets. To date, this work has been largely theoretical, although it draws on data gained from several Petawatt laser experiments in Europe. To test it, a high contrast ratio laser (i.e., one with virtually no prepulse) is required. This is counter to the normal design for the large laser fusion facilities such as NIF, where a prepulse is programmed to create an initial plasma that aids absorption of the main laser pulse. However, we are currently discussing doing some preliminary experiments at several other laser facilities where it is easier to implement the high contrast ratio requirement.

A second direction is the concept of creating voids in a host metal structure, which forms potential traps for formulation of ultra high density "clusters" of deuterium or deuterium-tritium. The clusters are a high density structure consisting of a thousand or so hydrogen atoms. When these hydrogen isotopes are diffused into the void-riddled metal, the clusters are formed. We have demonstrated that, but the next task is to create a structure that has sufficient numbers of clusters per unit volume to make an attractive target. Also, for an ICF target, the host structure must be restricted to low Z materials to prevent excessive radiation losses. Our work to date has employed palladium, so some fundamental changes would be needed to go on to low Z materials. The main advantage of this approach is that the target would be manufactured and injected at room temperature without requiring a cryogenic system such employed in all current ICF target designs. This enables use of a "frozen" D–T phase in the target, giving the high initial density desired for the implosion. Use of cryogenic targets would be a very difficult challenge for future power reactors, however. Targets must be injected on a Hertz time scale and enter through a very hostile environment which might prematurely vaporize the cryogenic fuel. Meanwhile, a sophisticated tracking system is needed to enable the pulsed laser beam to hit them in flight.

Our cluster concept which avoids cryogenics is quite preliminary, but it has received some interest in the target community. However, more recently we decided to concentrate on an alternate use of this unique cluster-filled material involving ion-driven fast ignition. The idea is to use it in a "convertor foil" (a thin foil filled with deuterium clusters) that is hit with a Petawatt laser pulse. The extremely intense laser pulse drives electrons out of the target forming a sheath of electrons in front of the foil. This sheath creates a MV potential which in turn draws out ions creating a MeV ion beam. This beam (deuterium in our case, where deuterium clusters were created in the convertor foil) is suitable for ion-driven fast ignition. The energetic deuteron beam is then directed into a precompressed ICF target. The energetic ions can penetrate into the center region of the ICF target, stopping there and depositing its energy on a small central volume or "hot spot." This hot spot undergoes a vigorous fusion burn which propagates outwards, igniting the surrounding fuel in the target. That is essentially the basic concept for "fast ignition." Fast ignition is of strong interest because creating the burn in the small volume hot spot requires much less input energy to burn the target than would be needed to compress the entire target fuel volume up to ignition. The hot spot approach then gives a much higher energy gain (fusion energy output over input energy) than the full volume heating case. The term "fast" comes from use of the ultra-fast Petawatt laser to initiate the hot spot burn. Fast ignition was first proposed by Max Tabak and colleagues at LLNL, and the concept is now being explored in laser fusion labs around the world. Several versions are possible, differing by the way energy is transported to the hot spot. Methods can range from use of electron beams or various ion beams using ions such as protons, deuterium, or carbon. In all cases, however, the particles transporting the energy are generated by use of an intense light beam from a Petawatt laser focused on a convertor foil or on the target itself. As will be discussed further in Chapter 12, our deuterium cluster converter plate for producing a MeV deuterium beam fits in well with ion-driven fast ignition research done on the TRIDENT Petawatt laser at LANL, and we have carried out encouraging preliminary experiments there.

The fate of my continued work on ICF probably depends on two factors: my continued good health and increased funding in the field, which could be inspired by a successful fusion burn at NIF.

## Further Reading

G. H. Miley, "Advanced-Fuel Targets for Beam Fusion" *Proc. of the 6th Intern. Conf. on High-Power Particle Beams* (*Beams '86*), Kobe, Japan, pp. 309–312 (1986).

G. H. Miley and H. Hora, *Edward Teller Medal Lectures,* World Scientific Press (2005).

G. H. Miley, H. Hora, B. Malekynia, and M. Ghoranneviss, "Reduction of Threshold for Laser Fusion Ignition by Nonlinear Force Driven Block Acceleration," *Fusion Science and Technology*, Vol. 56, No. 1, pp. 384–390 (2009).

G. H. Miley and X. Yang, "Advances in Proposed D-Cluster Inertial Confinement Fusion Target," *Journal of Physics: Conference Series, Sixth International Conference on Inertial Fusion Sciences and Applications*, Vol. 244, No. 3 (2010).

# Chapter 11

# Inertial Electrostatic Confinement (IEC) Fusion

*Left: This photograph shows me standing by our original IEC chamber in the late 1980s. The chamber itself is quite historic in that it was originally used by Joe Verdeyen in electrical engineering at the University of Illinois for measurements of noble gas plasma parameters in an IEC discharge. He obtained this unit from a contract granted through Bob Hirsch when Bob was at DOE. Bob had done the original IEC experiments with Philo Farnsworth in Fort Wayne, IN, in the 1970s and then moved on to DOE where he eventually became head of the Office of Fusion Energy. When I became interested in IECs in the 1980s (although I had done some earlier theoretical work in support of Verdeyen's experiments) I exchanged an optical bench for the chamber. As it turned out, we had to repair leaks in the chamber, convincing me that as usual Joe got the best of the deal. However, as it turns out we are still using the chamber in my IEC lab. Right: Kyoshi Yoshikawa and Masami Onishi from Kyoto and Kansai Universities are shown here with Dennis Beller from the Air Force Institute of Technology, Dayton, in front of my office door during some IEC discussions at Illinois. As described in the text, my interactions with Japanese scientists led to a number of exchanges and visits. Masami and his family spent two years at Illinois. Kyoshi spent a semester.*

Inertial Electrostatic Confinement (IEC) fusion has a fascinating history, having been invented by Philo Farnsworth, the inventor of the electronic television. The account of his life and the early days of television are documented in several books (such as D. Fisher and M. Fisher, *Tube: The Invention of the Television*, Publishers Group West, Washington, DC, 1996).

Farnsworth won his legal battles over the invention of the television, but became so far in debt financially in doing it that he could not produce any units. Farnsworth secluded himself on a farm in Maine where he conceived of the IEC concept for fusion. Remembering that Farnsworth was self-educated beyond high school, the concept, like his design for electronic television, is simply amazing. In his early patent on the IEC, Farnsworth provides a solution to the collisionless Valsov–Poisson equation in spherical coordinates. He found that the plasma formed a nested series of virtual anodes and cathodes around the center of the sphere. This resulted in increased ion densities in the trap regions between the virtual electrodes with the ion density increasing in each region going toward the center. At the center, the solution diverged with the density becoming infinite at a zero volume. In some ways, this configuration can be envisioned as an onion skin formation. Farnsworth termed the virtual structures "poissors." His mathematics treatment was much idealized, assuming among other things that the ions entering the device were monoenergetic and possessed zero angular momentum (i.e., with velocities directed along the radius and not at an angle to it). Those assumptions led to the many layer poissor solution. In reality, the many alternating virtual electrodes are reduced to two or so due to finite angular momentum and ion energy spread. Still, that configuration offers good confinement properties for the IEC. Farnsworth proposed that this configuration could be achieved in a small spherical vacuum vessel using a high voltage discharge to create the ions in fusing deuterium–tritium plasma.

It is difficult to comprehend how Farnsworth came up with the derivation of the poissors presented in his patent. The solutions he found for the Vlasov–Poisson equation are non-linear and the equation is tricky to solve. I have told people that after I read the patent, I decided to reproduce the solution to the equation. While I consider myself to be good at such mathematics, it turned out that my first attempts went astray and I did not come up with the full derivation for over a week. Yet, Farnsworth had not taken a university level differential equations course. Perhaps someone else helped him with the solution, but that has never been made clear.

Regardless, Farnsworth's ability to formulate the problem displays his genius. It appears that he was aware of two theoretical difficulties with electrostatic confinement that had to be overcome, which is exactly what he did. First, there is a well-known theorem by Earnshaw that states that plasma cannot be confined by electrostatic fields alone. This is easily envisioned physically, because an electrode placed in plasma with a voltage applied to repel one of the plasma species will end up attracting the other one, such that a plasma flow occurs and prevents confinement. This theorem, however, is based on steady-state conditions. Farnsworth's IEC concept was dynamic because the recirculation of the trapped ions provided an inertia (the "inertia" in IEC) that creates "stiffness" in the plasma. This in turn overcomes Earnshaw's theorem. The trapped, recirculating ions electrostatically confine the electrons, which are accelerated inward toward the center of the sphere forming the virtual cathodes. The second theoretical objection to this type of configuration, that Farnsworth was apparently aware of, came from an earlier paper by Elmore, Tuck, and Watson. In that paper, they considered electron injection into a spherical configuration such that the electrons are trapped, and in turn electrostatically confined ions. This is, in effect, a mirror image of Farnsworth's IEC where ions are injected. Elmore and his colleagues analyzed their configuration and concluded that the plasma density in the trap was limited by instabilities which would prevent interesting fusion power rates. Because their configuration seemed so similar to Farnsworth's, he could easily have been discouraged in thinking about IEC. Fortunately, he was not. The Elmore *et al.* analysis had failed to find the "poissor" type solution that Farnsworth discovered, so their conclusions were incomplete in that respect. In addition, the high mass of ions compared to electrons allows higher densities with their injection (versus electrons) before an instability sets up. The issue raised by Elmore and colleagues regarding electron-injected IECs has never been adequately addressed. I recently asked Rick Nebel, who at that time headed the electron-injected IEC program at $EMC^2$, about this. He responded that the original derivation made an error in the reference potential term. However, a revised analysis taking the potential term into account has not yet been published, so this remains an open issue.

I was not aware of Farnsworth's invention until I learned of it somewhat by accident. Farnsworth (then Vice President for International Telephone and Telegraph, IT&T) at the Farnsworth Research Laboratory in Fort Wayne, IN, called me to ask for a reference for Robert Hirsch, one of the first PhD graduates from Nuclear Engineering at the University of Illinois. Hirsch was excited about fusion, having completed a thesis with Ladislas Goldstein in the Gaseous Electronics Lab on a theta pinch plasma confinement device. (Early days in the Gaseous Electronics Lab were described in Chapter 4.) Farnsworth wanted to hire Hirsch to develop an experimental IEC program. I did not immediately realize who I was talking to, but responded quickly by saying that Hirsch was outstanding and recommended him without reservation. In retrospect, I later regretted that I did not engage Farnsworth in a longer conversation about the IEC when I had the opportunity.

The rest of the story is equally fascinating. Hirsch joined Farnsworth and carried out some amazing experiments on the IEC, producing remarkable fusion neutron levels from such a simple device. When told to approach the U.S. Office of Fusion Energy in ERDA (now DOE) to get more funding for this work, he took a small deuterium–tritium filled device with him to Washington to demonstrate to the fusion office personnel. When it was discovered that the device contained tritium, safety personnel prevented him from bringing it into the building. As a result, Hirsch did not win funding, but made such a strong impression that he was subsequently hired into the Office of Fusion Energy as a contract officer. At that time, the office supervised a small fusion research program with little funding. Due to retirements, Hirsch ended up as head of the office within a few years, becoming the youngest head of a government energy program at that time. In view of his interest in IECs, Hirsch administered several research contracts on the subject, including one at the University of Illinois, headed by Joe Verdeyen in Electrical Engineering. However, Hirsch had other plans. He recognized that the Tokamak program at Princeton University was slowly making good headway, so he went to Congress to seek funds for a vastly expanded "crash"-type program to push Tokamaks toward a fusion power plant within 20 years. His argument was that it was actually cheaper to do a crash program to develop fusion power as opposed to a long drawn-out program funded at low-levels which might never reach the objective of fusion power. His message got across and the Princeton Plasma Physics Laboratory was rapidly expanded along with Tokamak experiments at several laboratories in the United States.

A few years later, changes in the Congress resulted in a reduced funding level for fusion which was reclassified as a long term research program. Still the Tokamak program at Princeton, along with supporting work at other labs, notably Oak Ridge National Lab (ORNL) and General Atomics (GA) moved ahead. However, improved confinement in Tokamaks was found to require larger devices which became very expensive. Thus officials in the United States, Japan, Russia, and Europe agreed to move to combined funding of a single international Tokamak experiment for demonstration of reactor scale operation. This lead to an international project to develop a Thermonuclear demonstration reactor in France called the International Thermonuclear Experimental Reactor (ITER). However, ITER is very costly — over thirty billion dollars — yet it only represents a first step towards a true demonstration reactor which would produce electricity and confirm economics. With this time schedule, fusion power now seems to be a hundred years off! As noted under the chapter on ICF, one bright spot is that the National Ignition Facility (NIF) at LLNL expects to achieve a breakeven target "burn" with their laser fusion device in the near future. However, translating that physics achievement into a power plant will still require many years of effort to develop the necessary supporting technology. Also, it is not clear where the funding for technology development would come from. The Department of Defense (DOD), which funded NIF, is not charged with developing a power plant. The DOE Office of Fusion Energy would logically take that work on, but would need congressional approval for more funding. But, as already noted, most of their budget is concentrated on ITER.

The other way that fusion power plants might be introduced earlier is through a revival of funding for alternate confinement concepts as discussed here and in Chapter 8. Some of these concepts, e.g., FRCs and IECs, offer smaller and simpler components and size. Thus their development, assuming a physics roadblock does not occur as plasma conditions are scaled up, could proceed faster and with reduced cost compared to Tokamaks and ITER. But the DOE does not agree with this point, claiming that too little is known about the plasma physics of these systems because experiments to date have been quite limited.

After a few years Hirsch moved up in DOE to head all renewable energy research, including solar, wind, etc., in addition to their fusion work. This office was a political appointment, and thus Hirsch left DOE when the political parties controlling appointed offices changed several years after the expanded fusion program occurred. Since then he has held various high level positions in oil companies and is considered an expert on projections of future oil reserves. As a consultant for the RAND Corporation, he prepared a review of the U.S. fusion program for the National Academy of Engineering which was sharply critical of the obsession with Tokamaks to the exclusion of other devices. He publically stated that the Tokamak program was a mistake and that IECs, due to their simplicity and potential for smaller size and lower cost, were a better goal. This forced him into a political struggle with officers in the Office of Fusion Energy. As a result, he lost one of his consulting contracts for energy policy development. Meanwhile, there is virtually no government funding for research on IEC or other alternate confinement concepts. This brings out an interesting feature of our government. Due to the frequent changes in Congress, support for long-term research projects is jeopardized as politics and people come and go. However, military projects are somewhat shielded from scrutiny and rapid changes, so they are less likely to be cut prematurely. For example, the fission power program in the United States, while started by ERDA (former DOE), built its ultimate success on developments in the Navy's nuclear reactor program. Maybe history will repeat itself for fusion power?

There is a strong motivation to study IECs for near-term applications. It offers a practical neutron and charged particle source for a wide variety of applications, ranging from neutron activation analysis to production of radioisotopes for medical use. In addition there are credible predictions that the IEC can be developed into a fusion reactor. The reactor embodiment would be exceedingly attractive. It would be simpler than Tokamaks, eliminating heavy components like confining magnets, inductive current coils, etc. Also IECs can operate in a beam–beam like fusion mode (versus traditional Maxwellian fusion which occurs in traditional magnetic confinement systems). As a result, the IEC is well suited for burning advanced fuels like $D^3$–He, $^3$He–$^3$He and p–$^{11}$B. In an equilibrium system, the electrons are roughly the same temperature as the ions, and then cause excessive radiation losses in plasmas with higher Z ions like $^3$He and $^{11}$B. With the beam–beam type fusion possible in the IEC, non-equilibrium electrons have a much lower effective "temperature" than the ions, bringing down the radiation losses. Indeed, years ago the Electric Power Research Institute (EPRI) convened a board of scientists plus utility managers to review the potential role of fusion for electrical power from a utility point of view. Prior conceptual reactor design studies based on various confinement concepts including the IEC were reviewed. The advanced fuel IEC fusion power plant was selected as the most attractive plant. The report was sent to DOE, but the response was silence. Basically, staff there felt that this was an idealistic view of something that would never occur. Indeed, as has already been noted, the fate of many other "alternate confinement" concepts for fusion, such as the Field Reversed Configuration (FRC), Reversed Field Pinch (RFP), Spheromaks, etc., is that they too are simply ignored by the DOE. Their position is that the IEC (and these other devices) have never been experimentally tested at a scale close enough to breakeven to ensure there are no physics road blocks; whereas the Tokamaks had been studied in numerous experiments and were approaching breakeven in the last big experiments prior to ITER (e.g., TFTR, JET, and JT-60).

Thus, this becomes a "chicken and egg" situation. DOE's view is that the alternates have unknown physics while all such problems have been essentially resolved for Tokamaks. My counter-claim is that some key issues remain unresolved for Tokamaks, including ash buildup, alpha instabilities, and electric field effects. Bruno Coppi has another list of stability problems that he has loudly proclaimed for years. In addition engineering problems for Tokamaks like the high heat load requirements for diverters, neutron damage to the first wall components, and the large size required for adequate confinement translate into unfavorable reactor characteristics. All of this was summed up some years ago in Larry Lidsky's "White Elephant" article in the MIT technology journal as noted earlier. To address the size issue, researchers working on the various alternate approaches all try to reduce size by eliminating, to some extent, the use of bulky magnetic field coils. IECs go "whole hog" and use electrostatic fields instead. Many of the others substitute internal currents in the plasma to produce magnetic fields that lighten the field required from external current carrying coils. These changes bring in new physics issues, hence the challenge for researchers working on all of these concepts is to develop the theory and plasma physics to create a credible database for designing a breakeven demonstration unit. While these devices are typically smaller and much less expensive than an ITER, we were still talking about many tens of millions of dollars for such developments. Meanwhile, money for that level of support is not in the DOE budget!

As I finalized this book in 2012 the U.S. fusion budget situation had worsened. One budget bill under consideration in Congress would shut down remaining research experiments in the United States such as ALCATOR C at MIT in favor of increasing funding for ITER. This plan has received much criticism from the university research community.

Several years ago, while attending a DOE summer workshop on fusion in Aspen, CO, I had an interesting discussion with Jeff Friedberg, Professor at MIT and a well-known researcher on MHD instabilities in Tokamak plasmas. His comments bring out the view of many in the Tokamak community. After I gave a talk about the status of IEC development and an optimistic projection for going ahead to a breakeven experiment, Freiberg told me, "Great vision, George, but in Tokamaks, as we scaled-up experiments, new unexpected physics associated with various instabilities came in at almost every step. This required more experiments that had been anticipated and was a key factor leading to the long time required to get to ITER. When you and other IEC proponents try to scale-up towards breakeven, you will probably find the same thing happens. You are much too optimistic!" Well, maybe I am, but I like it that way!

Shortly after I restarted IEC research in the 1980s, Gerry Kulcinski, Professor and Head of the Fusion Technology Institute at the University of Wisconsin, asked me to attend and speak at the First D–$^3$He Fusion Conference that he was organizing. Wisconsin is the "home" for the $^3$He lunar mining concept. Jack Schmidt, a geologist, was an astronaut on the Apollo 17 mission who gathered the lunar "soil" samples that disclosed their $^3$He content. The $^3$He is the result of many centuries of bombardment of the lunar surface by the solar wind which carries $^3$He as a reaction product from the fusion reactions occurring in the sun. Unlike the Earth, the Moon does not have an atmosphere to absorb the $^3$He before it reaches the surface. The significance of this discovery initially escaped people, even Schmidt. However, he is an Affiliate Professor of Geology at Wisconsin, and discussions with Kulcinski and staff in the Fusion Technology Institute brought out that the lunar $^3$He could fuel D–$^3$He fusion reactors. Indeed, they estimated that the amount available from lunar mining would provide an energy equivalent that exceeds all the world oil reserves and more. Soon Lloyd Wittenberg, Gerry Kulcinski, John Santarius, and colleagues at Wisconsin published a landmark paper on this in the ANS journal, *Fusion Technology*, for which I served as editor. This article established the quantity of $^3$He available and made a case (perhaps controversial) for the favorable economics of mining and shuttling it back to earth to establish a D–$^3$He fusion power economy. $^3$He would then be the most valuable resource on the Moon and provide an economic incentive to establish an active colony there for mining operations. That vision is still quite vital, but as time moved on, the prospects for actually doing it remains uncertain. As I write this, we are not even sure when we will again have a manned mission to the Moon, largely due to the 2008 financial collapse. Regarding technical issues, however, things like the environmental impact (e.g., formation of an artificial atmosphere) of lunar mining need resolution. Equally important is the need to establish that a fusion reactor can be built to burn it. Tokamaks have almost no chance of doing this unless great improvements in their physics are found. In my opinion, the IEC would fit the bill perfectly!

Gerry Kulcinski knew my thoughts and that probably influenced him to invite me to speak at a workshop on $^3$He fusion at the University of Wisconsin. In fact, he was aware of our initial work on IECs, and was thinking of establishing a program at the University of Wisconsin (indeed, as time went on, they became one of the major research groups in the area). However, what I decided to speak on — use of the IEC as a small neutron source for industrial neutron activation analysis (NAA) — probably disappointed Gerry (at least at first, but much of the Wisconsin work has since been along those lines, greatly extending the concept to production of energetic charged fusion products for various applications like medical isotope production and $^3$He implantation in materials). Anyway, my talk on the neutron source concept was personally very eventful. John Sved from Daimler-Benz's German aerospace division was in the audience and immediately jumped on the idea. Ultimately, they licensed our technology and, using our help, constructed an "industrial" version for our IEC for use in NAA stations on ore belts entering their manufacturing plants. The IEC, with ~$10^8$ n/s from D–D fusion, replaced Cf-252 in these NAA applications. Cf-252 is a radioactive element that releases neutrons as it decays. It is produced "artificially" in a nuclear reactor and made available through the DOE. However, in recent years its production has been cut back, making supplies uncertain — yet due to decay the Cf-252 source must be replaced periodically. Even if Cf-252 is readily available, the IEC neutron source has advantages ranging from its favorable neutron energy distribution to its easy "on–off" capability. Placement of these IEC sources in Daimler plants was very successful, but John Sved's original hope to market units to others was canceled due to a downturn in revenues from the automobile division of Daimler-Benz. In any case, we claimed success and a landmark event (a Mute Swan?) of the first industrial use of a fusing plasma.

Why were we first? Compared to other possible fusion devices that might be considered as a neutron source, the IEC is compact, allowing a small size, and is the easiest to build and operate. The voltage applied to the grid directly accelerates ions. Thus, the IEC can produce 70-keV ions by applying ~80 kV to the grid, a relatively easy task. This ion energy corresponds to a high fusion cross section, allowing a high fusion reaction rate. Other fusion researchers would "give their right arms" to get such hot ions — for example, ITER strives for ~20 keV. In Tokamaks, the use of equilibrium (Maxwellian or Thermonuclear plasmas) results in much energy being wasted in heating of the electron population, which then transfers energy to ions via collisions.

We did add an important discovery to IEC technology by running with grids that have relatively large openings to create the "star mode." This mode creates ion beams that avoid damaging collisions with the grid wires by passing through the center of the openings. These beams are visible at higher background pressures, forming a beautiful image of a many-spoked star. One disappointment is that the star mode must run at higher pressures where most fusion comes from beam background neutral gas reactions which do not scale to a fusion power reactor. It is not that our vacuum pumps are poor. Rather, a reasonably high background pressure is needed to form the plasma discharge between the grid and vacuum chamber wall from which the beam ions are extracted and accelerated. I had purposely used this simple internal ion source design, rather than the external ion source technique like Hirsh had used years earlier. The internal source with star mode operation provides a simple inexpensive IEC neutron source. This approach has proved to be very successful.

I have always viewed this neutron source effort as a practical step along the way to an IEC fusion reactor. However, as time went on I came to realize it was only a "quarter" step — the physics is quite different from what is needed for a very low background pressure beam–beam reactor. Nevertheless, we did advance some reactor relevant aspects of the IEC in this step; I just wish it had been more. I remain fully confident that the reactor version can be achieved, but the road is challenging and requires considerable monetary support which does not presently exist. Our recent research on external plasma generation in a helicon for injection into an IEC for space propulsion is another step toward a reactor. This work is discussed in Chapter 14.

My IEC research at Illinois has received a tremendous boost over the years by a close collaboration with researchers from Japan. This connection originally started through a two year stay at Illinois by Masami Onishi, then Assistant Professor at Kyoto University. Masami was a close collaborator with Hiromu Momota who introduced us. In addition to the scientific aspects, Masami's family and mine became friends, and this led to enjoyable visits both ways over the years. Over time a series of five other Japanese researchers from various plasma groups there spent from 1/2 to 2 years in Illinois in our IEC lab. In addition, during this period an important series of IEC specialist meetings called the U.S.–Japan IEC Workshops were set up with financial help from DOE and JAERI.

These workshops have been attended over the years by a small group of key IEC researchers. Japanese attendees are typically from IEC groups at Tokyo Institute of

*Left: I'm shown with the Japanese delegation at an early US–Japan IEC workshop meeting during a party at my house. Hiromu Momota (second from right) later retired from Nagoya University and spent several years at the University of Illinois in the late 2000s. Masami Onishi is on the right side of Hiromu. Right: I was hosted by Kyoshi Yoshikawa and his wife (on my left) during a trip to Japan. Kyoshi created an IEC project at Kyoto University designed to use IEC neutrons for landmine detection. Later, he became a Vice President in the University and was on a Japanese National Planning Commission charged with restructuring the university research system in Japan.*

Technology, Kyoto, Kansai and Kyushu Universities and several industrial labs. American attendees have largely come from the University of Illinois, University of Wisconsin, University of Maryland, NASA Marshall, and LANL. Representatives from Australia and Korea have attended also. I hosted the first workshop in the series and they have alternated every year or so between Japanese and U.S. hosts. This series has been a very important factor in advancing knowledge in this field.

The IEC is simple, small, and can be used today to study fusing plasmas. Students realize this, and those inclined to work on laboratory-scale fusion experiments that they can get their "hands around" have joined the IEC work at the University of Illinois. I have been blessed with having a series of talented and dedicated students who worked on IEC experiments. Jon Nadler was instrumental in the beginning of the program, setting up our initial IEC device and getting things moving in his PhD work. John DeMora did combined computational and experimental studies of the optimization of the grid geometry for his MS. Brian Jurczyk did pioneering work in his PhD dissertation, which studied IEC physics in a variety of grid geometries. He and another PhD student at that time, Robert Stubbers, were instrumental in developing the IEC neutron source version of the IEC used by John Sved at Daimler-Benz. Robert Stubbers' thesis itself, though, was concerned with an alternate IEC geometry based on modified cylindrical type hollow cathode geometry. This configuration offers advantages by providing a rapid scanning capability for large area objects. After leaving the University and holding several jobs, Brian and Robert came back and founded a small company, Starfire Industries, located at the University of the Illinois Research Park. Starfire specializes in plasma applications, including neutron sources for applications such as NAA of impurities in coal at mines. Their source is not a traditional IEC, but came out of their experiences with IEC discharges. Starfire has been gradually growing, expanding their work areas and adding employees. A recent new hire, Mike Reilly, did his PhD thesis with me on a helicon plasma generator, spending much time at Edwards Air Force Base to take advantage of the facilities there for space trustier work.

*In this picture of the front face of the Starfire building in the University of Illinois Research Park I am shown with three former students. Brian Jurczyk is on the left, and Robert Stubbers is on the far right, along with Mike Reilly next to me. Brian and Robert founded Starfire and Brian is the president of the company. They have about a dozen employees. Much of their work grew out of their experience with the IEC in my lab. Although the present plasma products use somewhat different technology, Mike Reilly is one of their newer employees, having finished his thesis with me working on the helicon plasma source.*

Yibin Gu, who was in the Electrical and Computer Engineering Department, did his PhD with me on the IEC. Yibin developed a unique micro-channel collimator to study the source point and energy of energetic charge particles (protons) from D–D reactions in the IEC. He used the collimator data to infer the size and depth of the potential well formed in the IEC. This potential well must form to obtain improved ion confinement needed for power reactors. Yibin's thesis never got the attention in the IEC community that I thought it should, partly because others worried that this "indirect" type measure introduced uncertainties. Since then, others have tried various laser techniques, while one or two have used an improved (more accurate) collimator method like Yibin. The laser methods have ended up with large error bars while the collimator approach keeps improving. These workers, who now have personal experience with these very challenging problems, have come to appreciate Yibin's pioneering work more as time goes on.

Yibin took a job with AT&T and eventually ended up in the famous Bell Laboratory in New Jersey, where he combines his expertise from Electrical Engineering and Plasma Physics. One delightful experience came at his graduation ceremony on a beautiful May day at the Assembly Hall at the University. His whole family and some relatives came to cheer as he received his certificate on stage. Afterwards, we had some great discussions and fun. At graduation ceremonies, the PhD advisor has the privilege of joining officials on stage and handing the certificate to the graduate. However, few of my students have come back for the ceremony because it is usually months after they have left campus for jobs.

Another graduation ceremony that I vividly remember is Glen Sager's. His family also attended the ceremony and this had special meaning to me. Glen ended up at the University of Illinois because of an earlier discussion I had with his dad, Paul Sager. Paul worked on fusion engineering at General Atomics and was well known in the field. Some years back we had lunch together during a fusion meeting in San Diego. Paul asked me for help. Glen was about to graduate with a BS from UCSD as a top student with outstanding grades in math. However, he had become enamored with the wine vineyard industry in California and wanted to get a degree in that (offered in several California institutions). Paul wanted him to take advantage of his math ability, and in doing so, apply it to graduate studies in fusion. He asked me to help persuade Glen to do that! I did! Glen joined the IEC group and undertook a computational thesis. But then as Glen neared graduation, funding for fusion in the U.S. took a nosedive (characteristic of fluctuations in government programs). Thus it looked like Glen might not get a job in fusion research, whereas the wine industry continued to boom. I felt a great weight on my shoulders. As it turned out, however, Glen got a job at General Atomics just long enough to find an attractive opportunity at the University of California at San Diego Supercomputer Laboratory (founded by NSF, along with the University of Illinois's National Center for Supercomputer Applications (NCSA)). This job built on his experience gained in his computational thesis. I was off the hook! I could look at Glen's father and Glen in the face without flinching.

Another MS student in the IEC lab at this time was Ann Satsangi. Ann was raised on a farm in Minnesota, and she and her sister were star cross-country skiiers in high school. Ann went to the University of New Mexico for undergraduate work and made the U.S. cross-country Olympic ski team. However, an unfortunate leg injury prevented her from participating in the Olympics. The year she came to Illinois was one where we got little snow — a frustrating situation for a skier. When it finally did snow I suggested that she come over after work to practice on the golf course across the street from my house. There were a few times when the snow permitted this, allowing her to get a few hours of practice in before dark. One time I put my skis on and joined her, but when I got to the 9th hole she had already reached the 18th!

Ann took a job in the DOE office in Washington, DC, and has been very successful, now being in charge of the programs in high energy density laser physics and general plasma physics in the Office of Fusion Energy Sciences. Ann took the DOE job before she had completed her thesis work. I warned her that in my experience persons doing this often get so involved with work that their thesis is in danger of not being completed. Ann went ahead, and despite a real time crunch, she persisted and eventually completed her thesis — a real credit to her.

Several senior researchers and visiting faculty have made huge contributions to the IEC effort. Jon Nadler, as noted earlier, was the first PhD graduate from the IEC and took a very responsible job in the DOE operations office at Idaho Falls. However, after a few years, he returned to take a teaching position in Physics at the nearby Richland Community College in Decatur, IL, and also accepted a part-time research staff position on the IEC project. Jon was very experienced and really moved our IEC work ahead. And, as also described earlier, he took one of our devices to NASA's Marshall Space Center, where he stayed several months to help start the program there. This was a strategic time for establishing a role for the IEC in NASA's advanced propulsion program. Francis Thio (later the DOE-DOD Project Manager for high energy density physics) directed the advanced propulsion laboratory at Marshall. The IEC was one of four projects put into that lab with the intent of gradually upgrading each to larger power levels to evaluate their attractiveness for deep space mission propulsion. NASA scientists assigned to the IEC, under the leadership of Chris Dobson, along with Ivana Hrbud, then added more diagnostics and developed novel grid materials and grid manufacturing techniques. Progress was going well, but, unfortunately, NASA cut back most of its advanced research projects to save money for new shuttle missions (a big mistake, especially in view of the relatively small amount of money involved in these research projects relative to the shuttle program). It seems that in a funding crunch, research is always the first to be cut by many agencies and industrial organizations in order to meet near-term goals. But in many cases that has left them vulnerable over the longer term, as others move in with new technology. All four projects in the advanced propulsion lab at Marshall were gradually shut down. People left. Francis went to DOE as previously noted. Ivana Hrbud, who did the IEC grid developments, went to the Aerospace Engineering School at Purdue University. George Schmidt, who had spearheaded the advanced propulsion effort, went to an administrative position in NASA headquarters. (George is now the contact for the Office of Chief Technologist at NASA in Glenn, Cleveland, and is resuming his interest in advance propulsion.) John Cole retired from NASA Marshall. He had long managed projects on future propulsion research at all NASA labs. John and I still see each other at national aerospace meetings. We often spend time reminiscing about the "good old days" of advanced propulsion. He now reluctantly (because he has supported a variety of approaches) states that fusion propulsion is probably the best way to explore deep space. (Note the "probably" hedge — that's John!)

Over the years, a series of talented students have worked in the IEC group to tackle IEC physics with computations. Blair Bromley, initially working with me and then with Professor Roy Axford, an expert on advanced mathematical methods like Li Group Theory, did an extensive computer simulation study of the cylindrical hollow cathode IEC for his thesis. Blair graduated to return to his native Canada and is working on fission physics at the Chalk River Laboratory. However, he says he still follows the work going on in fusion. In addition to the hollow cathode experiments mentioned earlier, Robert Stubbers' thesis provided a detailed Monte Carlo analysis of this type of IEC. Later, Luis Chacon undertook development of a new computational method to study potential well formation in the IEC in collaboration with LANL scientists interested in IECs, including Rick Nebel, Dan Barnes, and Leaf Turner. The objective was to develop a self-consistent Fokker Planck treatment of the plasma confinement in the IEC potential well. This is a key IEC issue that came under question by Bill Nevins at LANL. Bill published a paper concluding that ion confinement in the potential well created in the IEC (without grids) was inadequate to give a reasonable fusion rate. His semi-analytic treatment seemed airtight, but did contain some subtle assumptions that left wiggle room. Indeed, after much work and an intensive simulation, Luis found that the IEC reactor premise survives if more realistic assumptions (e.g., a self-consistent potential well profile rather than a square well) are used. Luis's result was very encouraging to the IEC community — we could retain a hope for an attractive IEC power plant. As a side note, a second negative paper was published by Tim Rider, a PhD student of Larry Lidsky at MIT (author of the "Tokamak white elephant" paper). Rider concluded that the use of advanced fuels in IECs was unrealistic. Again, "the devil is in the subtle assumption made." In this instance, Rider used Maxwellian-averaged cross sections, rather than beam–beam averaged values.

Luis received a fellowship to work in the plasma theory group at LANL. This was a tribute to Luis' ability. While a small number of such fellowships are offered each year by LANL, this was the first in plasma physics for many years. His wife, Elizabeth (PhD in Civil Engineering from the University of Illinois) obtained a job there as well, and they remained there for some time. Recently Luis moved to a plasma theory group at ORNL. Unfortunately, from my selfish view point, Luis concentrates on toroidal plasmas now. But that is the top priority at the DOE labs. (As I was proofreading this book, I learned that Luis is now returning to the theory group at LANL!)

More recently, the character of my IEC work has changed, partly due to continued decreases in funding (in part due to the general demise of alternate confinement system support by DOE). We have moved to alternate versions of the IEC and alternate applications such as use in electric thruster systems. Rob Thomas from the Aerospace Engineering Department did a MS thesis on a unique dipole IEC concept intended for use as a space thruster. He showed that the insertion of a small dipole magnet could sharpen the beam focus in the central region, increasing the fusion reaction rate there. His presentations at several meetings gained attention. Linchun Wu did his thesis on heavy ion beam fusion, and in the process spent several summers at the Lawrence Berkeley Laboratory (LBL), the home for this approach under the leadership of my old LLNL mirror friend, Grant Logan. In addition to his very successful and widely recognized thesis work, Linchun spent time doing simulation studies of the IEC, partly in collaboration with Rick Nebel's IEC group at LANL (prior to Rick's taking over of the company $EMC^2$ as discussed later in Chapter 14). Linchun also employed the now famous (for ICF studies) LSP code developed by another former thesis student, Dale Welch. Dale was one of the students involved in my early ICF experiment as the University of Rochester as described earlier. He is now a principal scientist in a consulting firm in Albuquerque, NM, which specializes in computational methods development. LSP is a commercial code which is licensed to users for a fee. We worked out a "deal" where Linchun added some cross-section data in exchange for being able to use LSP. Dale is widely known in the field because LSP has gained extensive use at a number of laboratories. Dale also gave us some "free consulting time" on the IEC simulation, something his other users have praised him for doing. Linchun left Illinois after graduation to work on plasma simulations for a small company in Washington, DC. Even now, any time I have a question about China, or a translation question, I go to Linchun for help. For example before I attended the International Conference on Nuclear Energy (ICONE) in the fall of 2009, Linchun introduced me via email to several key professors at the University of Xi'an in China.

I mentioned Bruno Coppi earlier, but did not elaborate then. Indeed my association with Bruno has many memories worth noting here. He is an internationally recognized plasma physics researcher who left Italy to join the Physics and Fusion Institute at MIT. Bruno has recognized the virtues of using an ultra-high magnetic field to obtain a high plasma density, hence high fusion power density to obtain very compact fusion systems. He pioneered the ALCATOR experiment at MIT, which used the knowledge of people at the National Superconductor Laboratory at MIT to create superconducting coils for the ALCATOR. Bruno actively participated in all phases of the design and start-up of this very successful experimental device. An upgraded version remains at a vital facility today, an unusually long lifetime for fusion experiments! (Unfortunately, as I was finalizing this book, there was a danger that Congress would cut funding of ALCATOR along with other U.S. research to free up more funding for ITER.) As everyone in the field knows, Bruno is not only brilliant, but flamboyant in a characteristic Italian way. He has been an outspoken thorn in the side of Tokamak scientists with his many loud explanations of "flaws" in that physics. His Black Swan goal is to build a small high field ALCATOR-type experiment for fusion burn studies and then for a reactor. Through many loyal collaborators in Italy he has gone quite far in this direction without U.S. DOE endorsement. I had heard about this for years and thought it was dead! But, only recently, I learned that Russia is now seriously considering support of Bruno's project.

I had known Bruno for some years, but became close friends when he called me one day and said he had just read one of my papers and "seen the light" — the D–$^3$He Igniter was the only way to go!

While D–$^3$He fusion requires better confinement than D–T, Bruno calculated that the high field igniter was finally capable of meeting the requirements. The reduced neutron flux would be extremely beneficial in such an experiment, reducing induced radioactivity, simplifying shielding of the high field magnets, and reducing radioactive tritium concerns. Thus a leap-frogging of D–T to go to D–$^3$He in this experiment seemed very desirable and practical. I was delighted to have an ally like Bruno in advanced fuels. Bruno had stature and could generate excitement. In following discussions, Bruno said we needed to find a facility in the United States that had the power (motor generator set and incoming power lines) to run a high field D–$^3$He igniter experiment. I said, "Argonne National Laboratory (ANL) just mothballed an accelerator facility that had such capability." Subsequently Bruno joined me in a trip to ANL to propose this to managers in the accelerator division. Afterward they seemed to be persuaded to let us use the facility, but people in the Tokamak group (the main fusion project at ANL at the time) protested the possible drain of their money to this new project. After much debate, our proposal was ultimately turned down.

Bruno next invited me to join him on a trip to Italy where there was a science committee meeting in the Italian Congress to map out plans and funding for next step major projects. Bruno was to testify on the igniter and then he would yield the floor to me to explain use of D–$^3$He in it. We arrived late at night at the airport and flagged down a taxi. I was amazed when the cab driver recognized Bruno as a famous scientist. The hotel bellman and desk staff also recognized him! As it turned out, Bruno had been featured on several popular television shows discussing accomplishments of Italian scientists, both residing in and out of the country. Another coincidence was that the hotel made a mistake and overbooked rooms. Thus, I ended up sharing a room with Nobel laureate and past Director-General of European Nuclear Research Centre (CERN), Carlo Rubbia. He too had come to testify and request support for a new accelerator project. We exchanged some friendly words, but when he found I was with Bruno, relations quickly cooled. Bruno and Carlo were fierce competitors for Italian attention and funding. The rest of the time was a blur to me. All testimony and discussions were in Italian, which I scarcely understood. A translator repeated my English presentation in Italian to the audience. The translator seemed to be very excited and spoke so quickly I wondered if he might have added some of his own thoughts! Still, Bruno gave me a "thumbs up" when I sat down. After several hours of very heated debate, the Italian politicians went into a closed door meeting to finalize their decisions. When I questioned Bruno later about the outcome, he said that the recommendations were generally favorable, but still being negotiated. I got the impression that null-decisions are not uncommon in their Congress (like the U.S. Congress in recent years). Still, that was a great experience. This plus the ANL trip started a strong relationship with Bruno that continues to this day.

A low-voltage gridded IEC can be constructed relatively easily, so a number of amateur scientists have constructed and run such "fusors" (a name used by the website listed in the chapter references). The website provides detailed construction details for such devices, thus newspaper stories have appeared off and on talking about young "geniuses" who have built their own fusion devices. I knew that a number of such fusor home projects had been undertaken, but I was unprepared when in 2001 I received an email from Michael Sekora, a junior in high school in Pennsylvania at the time. He asked if I could give him a summer internship in my IEC group. Michael had built an IEC (as well as a home cleaning robot) for competition in a science fair in Pennsylvania, and had received a first place award. I was impressed, but was not set up to supervise high school students. Thus I thanked Michael for his interest and offered encouragement, but said I had no openings. However Michael would not accept my answer! After many emails and phone calls, I finally gave in. I wondered what I had gotten myself into when I later found that he was too young to live in a dorm and instead had to have a private room where his parents could give their permission for the rental agreement. I found a way to enroll him in a university summer camp program akin to youth music camps that are traditionally run in the summer at Illinois. That was essential in order for him to legally be in my lab (due to liability issues). However, what could have been a disaster, turned out to be just the opposite. Michael worked with the several undergraduates and grad students in the IEC lab exceeding well. He quickly caught on to the experiments and contributed much to the work. Plus he had a pleasing personality that allowed him to fit in well with the older students. I envisioned him coming to Illinois after he finished his senior year, but instead he went to MIT in the Physics Department. Before his first semester at MIT, he spent the summer working with my former student, Rick Nebel, on an IEC experiment at LANL. I wrote a glowing reference letter in support of his application there and, again, he more than lived up to expectations. I assumed he would work on fusion at MIT, but instead he moved into astrophysics. I followed his career there and found he was a top student in his class, president of the student American Physics Society branch, and had organized an intermural soccer team. Upon graduation, he went to graduate school at Princeton to continue work on theoretical astrophysics. I believe he will have a great career, and I like to think that the summer in my lab helped launch it, but with his talent, he would have done well no matter what!

One effort that I am currently working on is a book on IEC science and technology with S. K. Murali, as my co-author. Murali is a PhD who did his thesis work on the IEC at the University of Wisconsin. We hope this book, to be published by Springer Scientific Publications, will pull together information that can be used by persons wishing to enter the field as well as current researcher staff and students working on IEC.

## Further Reading

G. H. Miley, "Inertial Electrostatic Confinement for Neutron and Power Production," *16th IEEE/NPSS Symposium on Fusion Engineering*, IEEE No. 95CH35852, Piscataway, NJ, pp. 1419–1422 (1996).

G. Miley, B. Jurczyk, R. Stubbers and Y. Gu, "An Accelerated Beam-Plasma Neutron/Proton Source and Early Application of a Fusion Plasma," *Proceedings, 17th IAEA Fusion Energy Conference*, Yokohama, Japan, Oct. 19–24 (1998).

G. Miley and Y. Gu, "IEC Neutron Source Development and Potential Well Measurements," *Current Trend in International Fusion Research — Proceedings of the Second Symposium*, pp. 177–196, edited by E. Panarella, NRC Research Press, National Research Council of Canada, Ottawa, ON K1A OR6 (1999).

G. H. Miley, "New 'Spin-Off' Applications as Part of the Development Path for IEC Fusion," *Current Trends in International Fusion Research — Proceedings of the Third Symposium*, edited by E. Panarella, NRC Research Press, Ontario, Canada, pp. 547–558 (2002).

*The Open Source Fusor Research Consortium*, www.fusor.net.

P. Schatzkin, *The Boy Who Invented Television: A Story of Inspiration, Persistence, and Quiet Passion*, Tanglewood Books (2004).

# Chapter 12

# Low Energy Nuclear Reactions (LENR)

*Left: I visited Jim "Doc" Patterson (second from right) periodically at his lab in Clean Energy Technology, Inc. (CETI), a company he formed to develop and market practical cold fusion power cells. The CETI lab was located in a one-story metal building (once used for car repair) located near Doc's home in Sarasota, FL. His grandson, Jim Redding, standing by him here, had just graduated from business school at the University of Texas, and was the business and development manager for CETI. At the time, Larry Forsley was visiting from JFK Company in Washington, DC, and is shown on the left. Right: A number of the LENR meetings have been in Italy. The photo shows Zing Zhang Li, Bill Collis, Vittoria Violante, and me at one. Zing Zhang, Professor of Physics at Tsinghua University, Beijing, China, has been very active in LENR research over the years. Bill, originally from the UK, organized many of these meetings and is the operating officer for the Society on Condensed Matter Nuclear Science (CMNS) which hosts this series of meetings. Vittoria is a section leader in the Centro Ricerche Energia Frascati, Italy.*

I started out doing cold fusion research shortly after the infamous public announcement of "fusion in a test tube" by Pons and Fleischmann. However, I changed the terminology describing my work to Low Energy Nuclear Reaction (LENR) research. I felt this was more descriptive of the physics involved in my type of experiment which led to transmutation type reactions rather than D–D fusion sought by Pons and Fleischmann. This was done during discussions with John Bockus, a chaired Professor in Chemistry at Texas A&M. John was an internationally recognized electrochemist who became involved in cold fusion early on. He too had observed effects such as transmutations and in fact held two early meetings on that and other effects at College Station, TX. Those days and work are described in the interesting book *Nuclear Transmutation: The Reality of Cold Fusion* (T. Mizuno, Infinite Energy Press, 1998). The terminology LENR is now more widely used in the cold fusion community.

*Left: While not as frequent as the interactions I had with NPL scientists in Russia, I have had a number of interactions with Russians working on LENRs. Yuri Bazhutov (center) is shown here with me and Yasser Shaban. Yuri was the organizer of the annual Soviet conference on cold fusion and ball lightning held at the famous resort in Sochi-Dagomys on the Black Sea. He spent half a year in the U.S. working with me. During this time, he doggedly insisted that all my results could be explained in terms of "erizons," unique high energy particles he had discovered in earlier space physics studies. Right: John Fisher, retired physicist from the General Electric Research Laboratory, is shown here at a blackboard explaining his theory for LENR reactions that involves "poly-neutrons." He and Richard Oriani, retired Head of the Corrosion Chemistry Laboratory at the University of Minnesota, visited me in the early 2000s to explain how "poly-neutrons" might explain our LENR transmutation data. Richard was the experimental contributor to this team and they continue to this day to expand on the concept of poly-neutron reactions in LENR. This seems to tie into some features of transmutation reactions in LENR, but fails to explain other aspects. Thus, the jury remains out on this, but I have continued to have interesting and productive interactions with both John and Richard.*

The dramatic public announcement of "cold fusion" by Stanley Pons and Martin Fleischmann at the University of Utah occurred in March 1989. However, I first became aware of cold fusion a few days earlier when Steve Jones, Professor of Physics at BYU, phoned me. (I did not know it until later, but Jones was a competitor of Pons and Fleischmann. Their concerns about Jones' possible release of his work on the subject before they could disclose their work was an important factor in their decision to go ahead with the high profile premature public disclosure.) Jones and I had both worked on muon catalyzed fusion so we knew each other fairly well. Steve's phone call came just as I was preparing to leave my office for a trip to Japan. He asked me about the possibility of fast publication of an article on "cold fusion" in the journal *Fusion Technology*, for which I was editor. He told me that he was not sure if the topic fit into this American Nuclear Society (ANS) journal and wanted my decision on that. Plus, he was anxious to have rapid publication. I responded that I did not know what cold fusion was, and I had to leave for the flight to Japan immediately. I suggested that he send the article to me and I would start the review process when I returned in a week. I told Steve that if he felt it was an appropriate topic for the journal I was confident that it would be.

When I stepped out of the airplane in Tokyo, my host from the University of Tokyo met me waving a copy of the *Wall Street Journal* announcing the discovery of cold fusion. Thinking that I knew all about it coming from the U.S. where the discovery was announced my host anxiously asked me, "What is this cold fusion?" I couldn't respond, and thought to myself that it was too bad that I had not been able to take a few minutes to question Steve when he called me. As it turns out, the article Steve was speaking of was never sent to me, but was instead published through *Nature*. That publication played an interesting role in the controversy that grew between Steve and Pons/Fleischmann as cold fusion itself came under fire worldwide as replication of the experiment in many other labs failed. In a way, I am glad that I did not get to handle the paper — that would have drawn me even further into the controversy than happened otherwise.

The cold fusion announcement caused much excitement in nuclear laboratories around the world. Many raced to replicate the experiment. When I returned to my office from Japan, I again found a group of people waiting for me. This time they were students who wanted to do cold fusion experiments and hoped to get some heavy water ($D_2O$) from a supply that I had for experiments at the University of Illinois' TRIGA reactor. With my help, the students enthusiastically began several electrolysis experiments based on what was known of the Pons and Fleischmann device. They designed the experiment using my copy of the famous fax of an unauthorized copy of the Pons and Fleischmann paper they had prepared for a journal publication. This fax went from laboratory to laboratory around the world by fax machine. (Fax machines were a common means of communication in those days before wide use of the Internet.) The paper was nearly illegible from being transmitted so many times by the time I finally received it from a friend. Like so many others in numerous laboratories, our experiments did not succeed. Further, many rushing to do an experiment, including ourselves, failed to measure the amount of loading achieved in the experiment. It is now well known that success (achievement of "excess" heat) requires a loading ratio (atoms D/atoms Pd) over 0.85. Measurement of this ratio is difficult, typically done by observing the change in electrical conductivity of the electrode. Had the importance of the loading ratio been understood in these early days, many of the "failed" experiments would have been averted. At any rate, these experiences deeply influenced my interest in the field, something that I continue to work on today.

As already noted, many workers in the field have adopted the name LENR to replace cold fusion, because many reactions, strictly speaking, are not fusion per se, but instead are nuclear transmuted ions. Indeed, a more general terminology is Condensed Matter Nuclear Science (CMNS). But, turning back to the early days of cold fusion, shortly after the original announcement, I received an invitation to testify before a congressional committee investigating cold fusion in an attempt to determine whether or not a crash development program should be undertaken in the United States. By this time, however, a number of critics had come forth and the topic was emotionally debated in many circles. The congressional staffer who organized the congressional hearing wanted me to serve as an "unbiased" but open minded scientist recognized for innovative energy research. My testimony was placed between that of Martin Fleischmann and Harold Furth (then Director of the Princeton Plasma Physics Laboratory). Furth staunchly opposed cold fusion based on his hearing about failures to reproduce the phenomenon, plus his personal bias due to his long involvement in hot fusion. I served somewhat as a buffer between these two, but even so they waged an intense debate during the question session. Both were Brits and elegant speakers with quick minds, and armed with "cutting" remarks. My testimony was generally overlooked in the midst of all this turmoil. I mainly raised concerns about statements made by both Fleischmann and Furth. Luckily neither bothered to refute my comments, probably because they were more concerned about other politics involved in the situation. We were all exhausted and glad to reach the end of the congressional hearing and head for home. Martin Fleischmann passed away in August 2012 at his home in England. Some newspapers ran stories saying that this signaled the end of an era. But it might be said that the era of LENR that he was so instrumental in starting is just beginning. Before discussing that further, I should mention my involvement as an editor of several important professional journals.

In addition to doing research in the field, I was thrust into the controversy as editor of *Fusion Technology*. I began to receive papers that were submitted to the journal for publication, but by the time these started arriving opposition to the Pons–Fleischmann experiment had grown in the field. I had to make a decision about how to handle these papers. After much thought, I took the position that if they could get through our normal peer review process (three independent reviews) the paper should be published. A fundamental problem remained — opponents claimed that cold fusion did not exist, so any experiment making claims about the phenomena must be in error, and hence should not be published. In fact, editors of other major journals such as *Nature* and the various American Physical Society and American Physical Society publications took that attitude. Thus they simply returned any papers on the subject without submitting them for review. I took the view that the purpose of *Fusion Technology* was to disseminate information, so if reviewers could not find an error in an experiment or theory then the paper should be published. This then would provide serious studies and information to the broad scientific community, especially fusion scientists. As independent scientists they would have to decide what to do with said information (e.g., to ignore it or attempt to incorporate it into their research).

I published an editorial in the journal stating my position. As it turned out, I then became the only editor of a major journal who would accept cold fusion papers, and I received a flood of papers. Researchers in the field looked to me as their "hero." But some members of my editorial board were not happy. I met with this board at the summer and winter annual meetings of ANS. The board was in effect my "boss," so I served with their approval. They could, under extreme conditions, vote me out as editor. Many questions and comments on my policy for cold fusion papers came up at these meetings. For example, some demanded that I have at least two reviewers from the hot fusion community review these papers. I usually included at least one, but I had built up a list of researchers in the cold fusion community and often called on them. My argument was that hot fusion papers were reviewed by people in the field who were experts in the specific area of the paper, and the same procedure should, in my opinion, be applied to cold fusion. That argument satisfied the board for a while, but the subject kept coming up. Thus I started a policy of confidentially disclosing the names of reviewers plus copies of the review to editorial board members requesting this information for a specific cold fusion paper. I received a half dozen such requests and after providing this information about the reviewers the board members ended their objections. However, this issue was very divisive and the board remained split on their view of my editorial policy.

Some years later when I stepped down as editor (due to time requirements, not the cold fusion controversy), the first major change in the journal that was made by successor was to drop the cold fusion papers. Cold fusion researchers were frustrated, and to this day they have not found a "standard" journal as "home" for their papers. A journal has been created by the Condensed Matter Nuclear Science (CMNS) Society, which carries out a peer review process. It has not yet secured an "impact index" like the various professional society journals published by ANS, APS, IEEE, etc., so publications in it are not widely read. Now, in retrospect, I believe my policy on this issue was the proper one, reflecting the responsibilities of a scientific journal. I discussed some of this issue in an invited paper in the *Journal on Accountability in Research*, which is included in the references for this chapter.

I should add that the cold fusion episode was an important one for the journal and for me as an editor. To put things into perspective, I believe my major accomplishment was creating this journal (fusion papers were previously included in the ANS "sister" journal *Nuclear Technology*, with Roy Post as the editor) and building up its reputation in the fusion community. In addition to this journal, I am also proud of my role as editor of two major Cambridge Press journals, *Lasers and Particle Beams* and *Plasma Physics*. An editor's job is not easy. One must make the journal attractive to the field being served in order to attract outstanding papers. One factor involved in that is to have a reasonably short time between submission and publication. That in turn requires finding reviewers who can provide timely responses, both for the original paper and for a revised version if required. The occasional case where reviewers strongly disagree, such as when one recommends publication and the other recommends rejection, or when the author protests the review as being unfair, can be difficult for the editor. Diplomacy is needed to calm the situation but authority must be exacted to ensure high quality papers. All in all, the review process can be very time consuming and trying for the editor. In addition, journals are always under pressure to breakeven in terms of profit. Thus, the editor becomes involved in helping the publisher find subscribers. Despite the time demands, I found the job to be challenging but fun, and I am glad of my years devoted to being an editor. I feel good about doing this as an important service for the scientific community.

My approach to my own cold fusion research attempted to take a fresh look at the basic physics rather than simply duplicate the original Pons–Fleischmann experiment. While they used electrolysis to load deuterium into the palladium, I initially decided to capitalize on my plasma background and use a plasma loading technique. In order to obtain a high deuterium flux impingent on the target, I elected to use a pulsed plasma focus. In addition, discussions with Professor Heinz Hora led us to believe that an improved reaction rate would be achieved by using multi-layer thin film targets. The layers would be selected from metals that could form hydrides and also provide a large Fermi level potential difference at the interface between them. This mismatch in Fermi levels would cause a large local electron density that we reasoned would provide enhanced screening of the deuterons, leading to larger reaction rates. We termed this concept, "The Swimming Electron Layer (SEL) Theory."

To implement the SEL theory, we sputtered alternate layers of sub-micron thickness, nickel, and palladium onto a steel substrate. This appeared to be a promising approach, but unfortunately, the thin films fell off of the substrate after one or two plasma pulses. The combination of plasma heating (and possible nuclear heating) caused the films to delaminate due to their dissimilar coefficients of expansion. I reported these preliminary results at an early International Conference on Cold Fusion (ICCF) meeting in Hawaii. After my talk, Martin Fleischmann came up to me and said he thought I had a good approach, but needed to redesign the thin film electrode. Despite his encouragement, I became discouraged with this approach since the combination of the intense plasma flux and internal heating were disastrous for electrodes. Thus, I began mulling over ways to use such electrodes in an electrolytic cell. My whole view on this changed, however, when I went to the next ICCF meeting in Monte Carlo and met Jim "Doc" Patterson. Many people knew about Jim because his so-called "Patterson Cell" appeared to have the best energy output of any existing cold fusion experiment at that time. He formed a company, CETI, with his grandson, Jim Redding, as the business manager. They had an industrial booth at the meeting, which I happened to walk by. When I saw the electrode configuration of the Patterson Cell in their display, I realized that they were using small millimeter-sized plastic beads coated with metals as the electrode in a pack bed flowing electrolysis system. I caught Jim and said, "Those coated beads used for electrodes look like ICF fusion targets to me. I know a lot about such targets and think I could possibly make an improved version of yours." That "introduction" led to extended discussions, and we became good friends. Subsequently, Jim called me a few weeks later and suggested that he would bring one of his cells to Illinois. He proposed that I run his cell as an independent scientist to confirm his findings. I agreed, and later Jim brought the cell to Illinois and arranged for Dennis Cravens to visit us for a week to show us how to run it. Dennis was a consultant to Jim and was extremely accomplished at experimental electrolysis. The runs we made during that week were spectacular. The cell was designed to produce kilowatt outputs with only hundreds of watt inputs. The calorimetry for it employed thermocouples on the inlet and outlet of the tubes carrying the electrolyte fluid to the packed bed region. Even without the thermocouples, it was obvious that the cell was performing well due to some visual boiling and distinct spots where some electrode beads began to burn their way into the plastic wall of the vessel. Jim wanted me to develop an independent calorimeter to study this cell. I told him, however, that calorimetry was so controversial in this field that few would believe it no matter what. Thus I indicated that I would prefer to spend my time on something "more productive" and basic. I went on to propose a study designed to identify the reaction products using the extensive facilities at our Materials Research Laboratory (MRL) to do a quantitative analysis of the elements in the electrodes after operation. Jim agreed.

At that time, it was thought that the reaction products were from fusion reactions. However, Jim's cell used light water (i.e., $H_2O$ instead of $D_2O$) for the electrolyte so any fusion reactions would have involved protons (i.e., hydrogen or p–p fusion versus D–D fusion as envisioned in Pons and Fleischmann type electrolytic cells). However the p–p fusion reaction has an extremely small fusion cross section. Thus I could not understand how that could occur. I reasoned that the protons must instead react with the host-metal atoms. So, I decided that the first thing to do was to look for nuclear reaction products from p–Pd type nuclear reactions involving transmutations to other elements. The first step was to run a Secondary Ion Mass Spectrometer (SIMS) analysis on the electrodes before and after a run in order to measure and quantify any change in isotopes in the metal layers. This produced amazing results where a whole host of isotopes were detected in the samples following runs that were not present beforehand. Jim, Dennis, and I became extremely excited about this and that led to research on this effect that extended over the next four years. The obvious issue was a challenging one, namely to ensure that the measurements were not being confused by impurities in the system, or by artifacts such as molecular ions, etc., in the analysis. These issues were tackled on several fronts. We started by constructing a cell which used mostly plastic and glass parts to minimize impurities. For analysis I employed a new technique, which I termed "combination" SIMS and Neutron Activation Analysis (NAA). Because NAA involves measurement of distinct gammas with energies characteristic of the nuclear energy levels in an atom, confusion due to artifacts like molecular ions is avoided. Also, NAA can be done over a large volume, making it possible to do analysis of the entire cell and electrolyte to establish impurity levels in the total system before a run. One problem with NAA is that it can only identify elements that have a reasonable neutron reaction cross section. In this study, we were able to use NAA to identify seven key reaction products that had high neutron reaction cross sections and for which we had reference standards. I felt that their identification was essentially foolproof. Meanwhile, we used SIMS to roughly fill-in data for the other elements, normalizing the SIMS concentration data to the NAA measurements where that data overlapped the key NAA elements.

I first publically presented these results at a CETI advisory/consulting board meeting in Dallas, TX. One of the attendees, Peter Hagelstein, a leading theorist in cold fusion from MIT, challenged me for the lack of an NAA measurement of isotope ratios (many feel that if transmutation products occur, they should have a non-natural isotope ratio). The SIMS data did provide such information, but still had the uncertainties already noted. Thus NAA would carry the most weight. After discussions, I proposed we do NAA with two elements, copper and silver. The reason for this selection was that these had adequate neutron cross sections for such a measurement. We found a statistically significant deviation from natural abundance in the isotopes for both. Later, Peter came to Illinois along with an experimental nuclear physicist from MIT to review the data. After a lengthy discussion of the data, they agreed the transmutation data supporting the non-natural isotope ratios seemed convincing. That vote of confidence was very important to us because these two scientists had a reputation for being unbiased and extremely knowable and thorough. Despite this, Peter's theoretical development still focused on the "classical" D–D reaction experiments. He said the extension to transmutation type reactions would come later. Only now, after many years, has he started that extension.

Once we thought we had the non-natural isotope ratio effect pinned down, Jim Patterson, Gokul Narne (a student who did his MS thesis on this topic), and I published a full paper on these measurements. Gokul had spent many hours at the Materials Research Lab working on this analysis. His complete dedication to the work was a major reason we managed to do so much work on this so quickly. Also, Sheldon Landsburger, a Professor in NPRE, was at the time doing extensive NAA analysis on impurity metal transport in the atmosphere. Fortunately, the elements involved in his studies overlapped many we were concerned with, so his NAA setup was already calibrated against National Institute of Standards and Technology (NIST) standard samples for those elements. He then provided essential equipment and expertise for that part of our analysis. (When the TRIGA reactor at Illinois was shut down several years later, Sheldon moved to the University of Texas where he could continue NAA research plus other studies using their TRIGA. His loss, along with several others doing research at the Illinois TRIGA, was an unfortunate consequence for our department of the reactor shutdown.)

Gokul was an outstanding student and extremely enthusiastic about LENR as a result of his work for his master's thesis. I was confident that he would want to continue on to a PhD on this topic. However he came to my office one day to tell me that his father was quite ill and the family wanted him to return home to India to join his brother in taking over the family businesses. These businesses included a strange mixture of a shrimp factory and a coal-fired power plant. He felt obligated to return in order to help his family. Gokul did say that he would keep track of LENR research and try to do some research on the side. Despite his intent, that never happened. The business responsibilities consumed all of his time. Maybe this was for the best, as there are no jobs available in LENR, although that might be changing with the recent enthusiasm about LENR power units that I describe later.

I reported on these experiments at the ICCF meeting in Hokkaido, Japan. The magazine *Infinite Energy* traditionally covered ICCF meetings and highlighted the work as a "breakthrough." This publicity brought out all the critics of cold fusion, who were determined to attack the work in any way that they could. That was not unexpected. But, in addition, to my surprise many members of the cold fusion community itself would not accept that such reactions take place.

I was not the first to find such transmutation reactions. Two pioneers in cold fusion, John Bockris and Tadahiko Mizuno, had previously reported evidence for nuclear transmutations. Another researcher, Professor John Dash at Portland State University, was well known in the field of metallurgy and reported finding local regions where transmutations took place. But this earlier work had not received the attention that ours now received. The "hardcore" cold fusion group had ignored those earlier results, convinced that all cold fusion reactions resulted from D–D reactions, leading to Helium-4 (alpha particle) production along with lattice heating. Now my new results reinforced the older transmutation reports and provided quantitative data about reaction rates. This produced a strange situation. Once the original Pons–Fleischmann experiments came under attack, mainstream scientists in the physics and chemistry communities questioned all cold fusion work, saying that it defied fundamental laws of physics. Cold fusion researchers responded that these scientists did not have an open mind. However, now when something new came along in cold fusion, the cold fusion scientists themselves did not keep an open mind! As a result I could have been drawn into many arguments. My view was that this work stood on its own, and I tried to avoid loss of valuable time by getting caught up in these arguments, but at times that was not possible. But rather than debate endlessly, I wanted to continue new research to keep looking for the Black Swan.

The main reactions involved in the transmutation experiments were not, strictly speaking, fusion reactions. Rather many involved standard nuclear reactions or transmutations and some were a fission-type splitting of a heavy compound nucleus. The unusual aspect was that many of the reaction products could only be explained by assuming multibody reactions, e.g., multiple protons and Pd or Ni reacting as opposed to normal single-body p + Pd type reactions. Thus, as already noted, I adopted the name Low Energy Nuclear Reactions (LENR) for these studies and this terminology eventually gained wide usage in the field. Unfortunately the terminology "low energy" is ambiguous. Actually, what I had in mind was that the low kinetic energy associated with the reacting particles. Thus, unlike cases such as neutron induced fission where a large excess energy is involved, here the compound nucleus formed in the collision has very little excitation energy. Consequently, when it splits up, the reaction products are formed near ground state, so little radioactivity is involved. Thus, remarkably, a LENR reaction only leads to minor radioactivity; e.g., some soft x-rays and beta emissions versus hard gammas and long lived radioactive isotopes. The result then is quite different from fission reactions where energetic neutrons and gamma rays are emitted. In summary, these LENR reactions can be viewed as exhibiting two of the "miracles" cited against cold fusion by early critics: overcoming the Coulombic barrier without using high energy reacting particles, and yielding almost stable reaction products.

LENR, like hot fusion, requires considerable development before practical power sources are possible. Some workers in the field take the view that if LENR is developed, hot fusion will not be needed. That view puts the two into direct competition for the same R&D money, and pits cold and hot fusion scientists against each other. I personally view LENRs as an alternate approach to nuclear energy. However, I have always felt that both hot and cold fusion, once developed, will assume extremely important niches in the future energy economy. Perhaps cold fusion or LENR cells will provide small distributed power units while hot fusion will provide large centralized power stations. Thus both deserve vigorous research support.

Indeed, it appeared that Jim Patterson found his Black Swan with his Patterson Power Cell. Others agreed, and he was featured on the *Good Morning America* television show. However, several events occurred later that set Jim's work back. One was that Jim soon ran out of the batch of his electro-plated beads that worked so well. He had kept the production method in his head, and when he tried to reproduce it he was not able to make another batch that worked as well. This seems to be the fate of many cold fusion experiments — the experimentalist does not fully understand the key parameters that must be controlled, causing non-reproducible results.

I had been slowly working on an improved sputtered coating on the bead-type electrode, as I told Jim that I would do when we first met. My students and I looked at some of Jim's original electrodes under a scanning electron microscope and were struck by the disordered structures with cracking and defects. Intuitively that seemed to be bad for high levels of hydrogen loading in the metal coating. Thus I reasoned a "smoother" coating would enhance the cold fusion reactions. To achieve that, I envisioned using a sputtering technique to create smooth layers with good structure. This could be done by clamping single particles in a target holder in the sputtering chamber. However, we needed to make a thousand bead-type electrodes to go into one electrolytic cell. Thus, I eventually developed a "popcorn" shaker system that would keep the plastic beads suspended while sputtering onto them. The main worry with this approach was that collisions between beads would harm the coatings. We found that the harm would be minimal if the shaking frequency was properly selected. This process took about a year and a half to develop, but we ended up producing beautiful coatings (Mike Williams, who did much of work on the shaker unit, and I obtained a patent on the process.) Unfortunately, when these beads were used in the Patterson Cell, instead of increasing the performance, they caused a significant deterioration in heat production and transmutations. I was very frustrated. More recently I rethought this and decided that the sputtered electrodes eliminated the local defects in the films, especially at interfaces, where hydrogen or deuterium atoms could "cluster" together to create reaction sites. This is one of the factors that lead us to the deuterium cluster type LENR we are currently studying as described later in this chapter.

During the time when Jim Patterson was working on duplicating his original beads, his grandson, Jim Redding, suffered a fatal heart attack while exercising. This was a devastating event for Jim. His grandson primarily worked on the business side of CETI. But, in addition, he was Jim's constant companion and was also involved in many technical discussions and decisions. Jim Patterson never quite overcame the grief of losing his grandson and fellow worker. While he continued work in his lab, it was never the same after that. A few years later, in 2008, Jim Patterson too passed away. This family team in CETI spanning two generations was an unusual and memorable part of the turbulent history of cold fusion.

My transmutation results caught the eye of Lewis Larsen, who called me to introduce himself, saying that he heard of this work. Lewis lives in Chicago and proposed a visit to discuss the possible development of a company to create and commercialize these cells. Lewis has a degree in biology but was largely engaged in financial management and stock trading in Chicago. He was very quick in grasping a wide range of topics and had a real knack for making contacts with important people. After several meetings, Lewis set up a company, Lattice Energy, LLC, with me as the Chief Scientist and himself as CEO. Money was raised to support an aggressive program to develop a heat producing cell based on my thin film electrode design. A patent issued to me on the subject was the first major intellectual property owned by Lattice Energy. This association and support lasted several years until Lewis and I found we could not agree on plans and views. Basically, Lewis became convinced that a commercial unit was possible in a much shorter time frame than I did. Also, I favored more open publications than Lewis. We parted on agreeable terms and remain friends, though we do not meet often. Lewis continues to push LENR, supporting several other projects through Lattice Energy, LLC. Most importantly, recently Lewis and Allan Widom, a physicist at Northeastern University, published a series of theoretical papers on LENR called the "Widom–Larsen Theory." This theory, involving formation of a heavy electron, which leads to low energy neutrons, has been quite controversial in the community. However, one interesting fact is that it does predict the array of transmutation products I observed in the Patterson Cell experiments reasonably well. In this theory, the low energy neutrons create the transmutation reactions so issues of how charged ions could overcome the Coulombic barrier are circumvented. Scientists at NASA Langley are attempting some basic experiments to verify this theory, but no results have been released yet. In any case, I proudly add Lewis to the list of people I have introduced to the field of LENR to the extent that they have become active in it.

I wrote this book in sections. One area in the LENR discussion that I put off in order to pay particular attention to it was my lengthy association with Russian scientist Andrei Lipson. But events overtook me just as I finally started that section, Andrei unexpectedly died in late 2010, suffering a heart attack while riding a subway train in Moscow. He was only 50, and I had just seen him several weeks earlier at the ICCF-15 meeting in Rome. He appeared vigorous at the meeting, giving talks and entering many side discussions. He seemed to be in good physical condition except for his continued smoking. His death was most unexpected and tragic. There were many e-mails on the Condensed Matter Nuclear Science (CMNS) server from colleagues around the world expressing their shock and sympathy. Andrei was a well-liked and highly respected member of the LENR research community. He left behind his wife and daughter. Indeed, his daughter, Maria, attended the University of Illinois studying art history. She ended up at Illinois as the result of the family accompanying Andrei to Illinois over a decade ago, when he first came to collaborate with me. Although Andrei and his wife returned to Moscow several years ago, Maria remained in Illinois to continue her studies. The loss of her father has been hard on Maria. He had wanted Maria to go on to graduate school, but lack of financial support after Andrei's death made that questionable. Somewhat by luck, I learned of a student job opening during clerical work in the Aero Department and recommended Maria for the position.

*Andrei Lipson (second from left) with colleagues Carlos Sanchez (on the left) and Tadahiko Mizuno and Xing Zhong Li (to the right of Andrei) at the ICCF 15 conference in Rome shortly before his death. As discussed in the text, Andrei Lipson was from the Institute of Physical Chemistry and Electrochemistry at the Russian Academy of Sciences, and he spent four years working with me. (Photo by David Nagel)*

She got the job, and it enabled her to complete her senior year. Then, after applying for graduate work, Maria fortunately obtained one of only two TA positions available in her department. Liz and I recently attended her graduation ceremony. She was one of four students in Art History to obtain a MS. Maria plans to go on to a PhD and hopes to become a curator in a major museum. Andrei would have been so proud of her!

Andrei was a great colleague who had a significant influence on my LENR research. It seems like only yesterday that I first met him at an ICCF meeting in Europe. He had just returned from a stay in Japan where he collaborated with Professor Kosagi on low-energy nuclear cross sections using ion beam–target experiments. I approached Andrei after his talk and asked if he would consider coming to Illinois to join my work, assuming I could raise enough money. He consented. Later, in e-mails, I asked if he "believed" in cold fusion. He replied that he thought so but wanted to keep an open mind because the experiments would determine the truth. I told him that I shared that view. From that moment on we usually agreed on overall goals, though we differed in a professional way on some details. Over his four year stay with me, punctuated by periodic visits back to his lab in Moscow, Andrei and I formed a very productive team. Not only did we do continued experiments on thin-film electrode concepts, we successfully applied some analysis techniques that Andrei used in Russia, such as nuclear particle detection using CR-39 film. We also developed in collaboration with others at Illinois a pioneering method for creating near metallic density hydrogen (or deuterium) states in dislocation loops in palladium. The resulting *Physics Review* article about the superconducting properties of this state has received a number of citations.

This work with Andrei culminated in my most recent LENR power cell concepts based on deuterium "cluster" formation and reactions. We were working on this concept when lack of support forced Andrei to return to Moscow. He continued to collaborate with us on cluster research working from his lab in Moscow. Ironically, several weeks after his death, we received a package from him postmarked the day he died. It contained a new cluster type electrode he formulated for us to study. He had mailed it on the morning before his fatal heart attack! In addition to new experiments at Illinois, Andrei brought with him a wealth of data from experiments in Moscow that had only been written up as obscure lab reports in Russian. I helped Andrei provide added theoretical understanding for select experiments (ones that seemed to have the highest scientific importance) and worked to develop an English version of the Russian reports. Over six publications in technical journals resulted in this manner. Subjects ranged from pyrolytic fusion neutron production to LENR transmutation studies. One old publication by Andrei about "bubble fusion" was already in English, but was often overlooked. Based on that report, it appears that he and his group in Moscow were the first to achieve neutron production using the cavitation technique. Thus, when the article in *Science* by Rusi Taleyarkhan from ORNL came out about bubble fusion without mentioning Andrei's early work, Andrei was very upset. I proposed that he submit a letter to the editor, which he ultimately did. However, apparently due to the growing controversy over the original article, Andrei's letter was not printed.

There is not enough space to tell all I want to about Andrei's accomplishments and our close relationship. I grew to deeply respect him as a person and as a scientist. The interested reader can find out more about him from the article by his daughter, Maria, in *Infinite Energy* magazine following his death. Also, I spoke about some of his recent research at the March 2010 American Chemical Society meeting in San Francisco, CA. Andrei had prepared a presentation on the effect of electron beam bombardment on loaded hydrides for that meeting. I attended the meeting and made the oral presentation that had been scheduled for Andrei. He had sent a draft PowerPoint presentation to me for comment just before his death. Thus I was able to use many of his own slides for that talk. It was very sad that he was not with us to present it.

Another person who has had a very strong influence on my LENR work is Charles Entenmann. He retired from his family bakery business and now lives in Florida. Somewhat by accident he met Jim Patterson and became a collaborator in the CETI cold fusion work. Jim introduced me to Charles, who became a friend and strong supporter of my LENR research. He, along with his son-in-law, Alf Thompson, accompanied Jim for several visits to Illinois to discuss LENR with me. We also got together in Florida during meetings at CETI. Alf was particularly interested in transmutations and was collaborating on an experiment when, in early 2011, he died unexpectedly from a stroke. This was a blow to all of us. Charles and I recently wrote an article for the *New Energy Times* in Alf's memory.

At this point, like some others, the reader may find it unusual that three relatively young people in LENR (Jim Redding, Andrei Lipson, and Alf Thompson) died over such a short timespan. Possible causes including strange radiation from LENR cells to a conspiracy of some type have been discussed by persons on the Internet. Personally I feel this is just an unrelated coincidence and that there is no relation between these unfortunate deaths. I hope to provide a counterexample.

Xiaoling Yang, a post-doc with me, has played a leading role in our present deuterium cluster LENR research. She and a group of four enthusiastic undergrads have done most of our LENR experiments over the last two years. Xiaoling's post-doc position was to end in May 2013, but she obtained an excellent full-time position in a biochem company on the East Coast and left for that in December 2012. Xiaoling is hard to replace. Heinz Hora continues to provide valuable support in the theory of these unique clusters. Nie Luo also makes important contributions to the work, but is limited in time due to his other projects. Charles Entenmann continues to participate in email and phone discussions. An interesting and important aspect of the work is the connection to ICF fusion. We have proposed using the same cluster type foils used in LENR cells for a convertor foil to create energetic deuteron beams for fast ignition of ICF targets. Fast ignition involves hitting the compressed target with a burst of focused energy from a laser or ion beam to ignite a hot spot (small region of burning fusion fuel embedded in the larger volume of fuel) to achieve a target burn with minimum energy input. We have been collaborating with Kirk Flippo at LANL to study this possibility experimentally using the TRIDENT Petawatt laser facility. Kirk and others at LANL have been studying use of energetic proton beams for this purpose. Deuterium is similar, but has the advantage of providing added fusion by interacting with the D–T target to create added fusion reactions while it simultaneously creates heating of the hot spot. Before our cluster-type foils, there had not been a good way to make the convertor foil that the laser hits to drive out and accelerate the deuterium ions. Xiaoling took several "cluster" foils to LANL for preliminary experiments which turned out to be encouraging. Thus, it looks like clusters could provide a scientific "tie" between hot and cold fusion!

Shortly after I finished the initial draft for this chapter, the LENR situation changed drastically. Andrea Rossi in Italy announced that he had developed a commercial scale 1-MW unit which could be purchased, and announced dates and times for upcoming demonstrations of its operation. This announcement resulted in much excitement; however the excitement was tempered by debates about whether or not Rossi's work could be taken seriously. These debates were intensified by Rossi's refusal to release full details about the unit, and the restricted viewing arranged for the demonstrations. Rossi defended his stance, saying that as a private company he needed to maintain proprietary information. However, he did disclose that the unit was based on a hydrogen/nickel reaction using nickel nanoparticles along with a secret "catalyst." Other information slowly came out, including the observation that nuclear transmutations were occurring in the metallic nanoparticles, resulting in production of significant amounts of certain elements in the nickel such as copper after extended runs. There is not space here to provide more information about the strange events that occurred and continue to occur around Rossi's activities. Much of that story can be easily found from an online search.

When I first heard about Rossi's work, my initial thought was that this reminded me in many ways of the Patterson Cell. Patterson too claimed high power outputs, but maintained secrecy, including restricted demonstrations due to his company CETI's plan to commercial the cell. That also immediately led to controversy about his work. Jim used light water and nickel in the Patterson Cell, so one would think that the reaction involved would be quite similar to the Rossi hydrogen gas-loaded nickel cell. Further, when I studied the reaction products from a Patterson Cell, as already noted, I found that nuclear transmutations in the nickel produced a variety of elements, including copper. Rossi too reported finding copper in his electrode after a run. Thus I began to think that I should get back into this area to take advantage of my considerable previous experience with such systems. In addition, I had always been impressed with the earlier work on gas loaded nanoparticles done in Japan by scientists such as Arata and Takahashi. These experiments also produced very interesting results, but at lower power levels. I did not worry about power levels, because looking at it from a scientific point of view, all the basics can be learned from low-power laboratory experiments. But outside the scientific community, low powers do not attract much attention, whereas discussion of a commercial MW unit does! Thus at the next meeting I went to where LENR was discussed, speakers and the audience focused so much on the Rossi work and speculations about it that other good scientific work received little attention.

At any rate, I set up a small gas loaded nanoparticle experiment in my lab, and assigned Xiaoling Yang and several undergraduates to work on it. Xiaoling had a great background for these experiments. She was already working on hydrogen/deuterium clusters in thin foils, and this could easily transition into nanoparticles. Xiaoling and I reviewed our previous thin film work with light water and nickel/palladium films as a basis for selecting the composition of the nanoparticles. We were able to obtain the desired alloys, and milled the block of alloy into nanoparticles using facilities at the Materials Research Lab at the University of Illinois. The experiments went rapidly and within months we had some exciting positive results. I reported these results at several meetings with sessions on green energy, and also spoke at an invitation-only meeting set up at NASA Glenn Laboratory in Cleveland. At this time Rossi's work was still under debate. It created such excitement that several companies and also NASA scientists wanted to obtain a Rossi-type cell or obtain detailed information about operation of such a unit. Simply put, they were not sure that this was going to turn out to be practical, but did not dare ignore it, because a LENR reactor with large output energy and negligible radioactivity would be a game-changing power source. Rossi, however, turned out to be very difficult to work with, and refused to sell small demonstration cells for this purpose. Meanwhile, although I thought we were late getting into the game, the results from our experiments seemed to be the best to date in the United States (some others may be keeping their work secret, so the exact status is hard to know). Anyway, this resulted in a number of individuals and companies contacting me to either obtain access to our work or with offers to collaborate in formation of a company to commercialize it.

As I composed this update on LENR, I had just completed negotiations to set up a new company which would be a joint venture for NPL Associates, Inc., called LENUCO LLC. If angel or venture capitalist funding can be obtained we plan to move the LENR work into a laboratory in the University's research park. One major technical challenge remains that I feel we must solve before moving our university research to LENUCO. Currently run times are limited by "sintering" of the hot nanoparticles. To be of comercial interest, run times lasting many months before replacing the nanoparticles seems essential. We are currently examining some techniques to accomplish that, and hopefully we will be successful soon. The objective is to move our technology ahead as rapidly as possible, so that it could be commercialized before too many competitors catch up. Everyone involved realizes that this is a risky endeavor, because there are still many issues to be resolved before we can be sure that a competitive product is possible. Still, the possibility that this is likely to become a game-changing power technology is enough to excite and drive us all ahead.

## Further Reading

G. H. Miley, "Quantitative Observation of Transmutation Products Occurring in Thin-Film Coated Microspheres during Electrolysis," *Progress in New Hydrogen Energy*, edited by N. Okamoto, Vol. 2, p. 629 (1997).

G. H. Miley, "Some Personal Reflections on Scientific Ethics and the Cold Fusion Episode," *Journal on Accountability in Research: Policies and Quality Assurance*, Vol. 8, No. 1, (2000).

G. H. Miley and H. Hora, "Maruhn-Greiner Maximum of Uranium Fission for Confirmation of Low Energy Nuclear Reactions (LENRs) via a Compound Nucleus with Double Magic Numbers," *J. Fusion Energy,* Vol. 26, No. 4, pp. 349–355 (2007).

G. H. Miley, H. Hora, K. Philberth, A. Lipson, and P. J. Shrestha, "Radiochemical Comparisons of Low Energy Nuclear Reactions and Uranium," *ACS LENR Sourcebook* (2009).

G. H. Miley, X. Yang, and H. Hora, "Ultra-High Density Deuteron-Cluster Electrode for Low Energy Nuclear Reactions," *Proceedings of American Chemical Society* (2010).

G. H. Miley, X. Yang, E. Ziehm, and H. Hora, "A Space Power Source using Low Energy Nuclear Reactions (LENRs)," *AIAA*, Reston, VA, and *IECEC Conference*, Atlanta, GA (2012).

V. Violante, F. Sarto, E. Santoro, L. Capobianco, M. McKubre, F. Tanzella, G.H. Miley, N. Luo and P.J. Shrestha, "Study of Lattice Potentials on Low Energy Nuclear Processes in Condensed Matter," *ICCF10*, Cambridge, MA, Aug 24–29 (2003).

# Chapter 13

# Hydrogen Economy and Fuel Cells

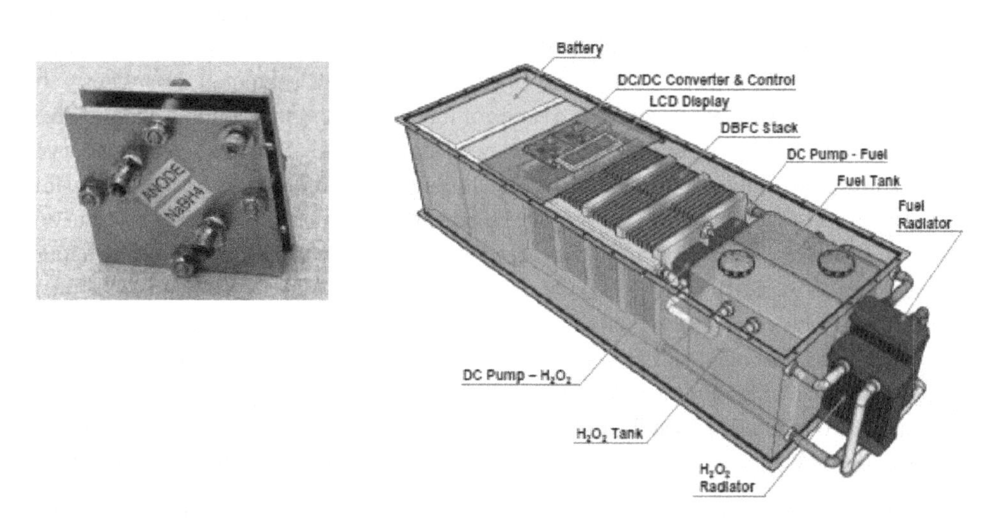


*The small watt level all liquid fuel cell on the left was the first one built in my lab. While only producing watts, it provided the test vehicle for perfecting catalysts and other components to achieve very high power densities using our novel all-liquid-type fuel cell. This then allowed us to design and build very competitive stacks (multiple cells coupled electrically in series) for major applications. As an example, the design on the right shows a 1-kW direct liquid fuel cell stack for use in an Unmanned Underwater Vehicle (UUV). These Direct Borohydride/Peroxide Fuel Cells (DBFC) are ideal for applications like this where the oxygen comes from hydrogen peroxide, making them air independent. We came into the fuel cell area late, but major grants from NASA and DARPA allowed us to very quickly become a leader in this specialized type of fuel cell. Two talented research staff, Nie Luo and Kyu-Jung Kim, as well as several outstanding students, have worked on this project. On the teaching side I have, with much help from Nie, Kyu-Jung, and Professor Magdi Ragheb, introduced two new courses on "The Hydrogen Economy and Fuel Cells" and on "Energy Storage" that fill a gap in offerings in the College of Engineering at the University of Illinois. This is quite a departure from the plasma and fusion courses I have traditionally taught. However, I really enjoyed this opportunity.*

My original undergraduate education was in Chemical Engineering with a minor in Physics, so it might seem natural for me to get into fuel cell research. However, having worked on nuclear energy for over 40 years, I had not been thinking about fuel cells or the hydrogen economy until recently. In 2004 a series of events occurred that changed this. Nie Luo, who had come to work with me as a post-doc on LENR research, brought up the subject in a discussion about cold fusion electrolytic cells. He had some thoughts about using energetic fuels like hydrogen peroxide and borohydride in a fuel cell. While the discussion was interesting, I had almost forgotten about it. Then one day Nie pointed out a DARPA SBIR request for proposals that included a topic on high powered fuel cells for military applications.

The opportunity to apply our newfound electrolytic knowledge gained from LENR fuel cell research was appealing. An all-liquid fuel cell of the type we had envisioned would offer a very high power density, making it of strong interest to DARPA. Nie quickly carried out some exploratory experiments to prove our concept and we submitted the proposal. It was ultimately funded, launching us on a series of fuel cell studies that continue today. Mike Obal, the DARPA project manager, was very knowledgeable and also provided the guidance needed to successfully negotiate this project through all three phases for a successful DARPA project. The last phase involved development of a demonstration cell for use in an Air Force satellite. The work was quite successful and led to several other interesting applications. The most recent involved construction of a small 70 W unit for Sandia National Laboratories for use in a robotic "urban hopper." This urban hopper looks like a small model tank. It is radio controlled and can be employed in a battlefield situation where it can leap many feet into a window. Our fuel cell was viewed as an advanced power unit versus the standard Li-ion batteries used in these hoppers. Tests with our unit went well, but the demand for this type of urban hopper dropped, so only a few of our fuel cells were employed.

*Nie Luo, in the foreground, and I are shown in our machine shop with a 1-kW fuel cell on the drill press stand. Nie joined my lab after graduating from Northwestern University with a PhD in theoretical solid-state physics. But it turns out he is equally talented in the lab and can "make things work." He and Kyu-Jung Kim, who came from a management position in LG corporation of Korea to work with us, have been instrumental in this fuel cell effort. Nie has also strongly contributed to other projects including LENR and plasma research.*

Doing fuel cell research in our NPRE department has some drawbacks. NPRE students do not normally feel they have a background for such work, so are often reluctant to join our research group. Thus, I have frequently involved students from other departments. One close association sprung up when we obtained NASA granted support for our fuel cell development based on a Broad Agency Announcement (BAA). Using peroxide frees our cell from the need for air to supply oxygen. Thus, it is a natural for use in space. Rod Burton, Professor of Aerospace Engineering, originally pointed out this BAA opportunity to me. Later he enthusiastically joined in this work with us and was an important contributor. Some years earlier I was appointed as an Affiliate Professor in the Aerospace Engineering Department due to my interest in fusion propulsion. This matched well with Burton's research on electric propulsion. Now we also share a common interest in fuel cells. While the fuel cell was not for propulsion, fuel cells have been used for energy sources in numerous prior NASA space missions, so Aero students "relate" to them. This association brought in students from Aero and allowed us to set up a fuel cell lab in the Aero Laboratory area to supplement mine in the Nuclear Engineering Laboratory. This NASA work was very productive, but was canceled when NASA decided, after a year into the project, to cut some 80 "advanced" BAA projects nationwide to concentrate on near-term issues. Fortunately, they agreed to extend student support (but not other costs) for a year to "ease the blow."

I must note here that sadly this cancelation of advanced research happens all too frequently in U.S. government agencies. As Congress changes, so do the priorities and budgets. This causes a "spicket on–spicket off" mentality. This has held back long-term research in the U.S. in recent decades. As I am writing this, the pendulum has again turned somewhat. The 2010 NASA budget seemed destined to drop near-term manned missions and stress to some extent longer-term research to get ready for aggressive future missions. The question is: how long will it take to rebuild resources, especially the skilled researchers needed to get it going again?

Nie, Kyu-Jung, and I remain leaders in the United States in borohydride-peroxide fuel cell research. However, to date this type of cell has not received wide interest, mainly due to concerns at DOE that reprocessing the waste boroxide (formed from borohydride oxidation during operation) back into borohydride is too expensive for wide use, such as in automobiles. This would be done at a base chemical plant, and companies involved in the area, such as Rohm and Haas Chemical (now part of Dow Chemical) state that the costs would come down drastically once a sufficient market develops. Thus, the cost issue becomes a "chicken and egg" problem because the market will not expand until the cost comes down. DOE has, however, concluded that a related fuel, ammonia borane, may stand a chance of economical recycle. Our all liquid fuel cell technology can be converted to this fuel, but even so that has yet to impress DOE. Basically, they have set their sights on hydrogen-oxygen gas-based systems, largely because they have already invested so much money in such studies. Indeed, some demonstration hydrogen gas stations using high pressure tanks for hydrogen storage already exist in select locations across the country. In retrospect, it seems that this all-liquid fuel cell plus other areas of my work like IEC fusion end up outside of DOE's mainline programs, greatly reducing chances of government funding from them. In a few instances I have tried to take the view, if you can't convince them, join them. Hence in IEC fusion, I added a dipole focusing magnet. This has technical merit because the fusion plasma density can be enhanced. However, an ulterior motive was to bring a magnet in, because DOE concentrates so heavily on magnetic confinement fusion. That strategy worked initially, but the DOE funding for the dipole IEC dried up later, as did funding for all of the alternate magnetic confinement approaches due to the emphasis on the international project ITER. While focus and funding in the United States narrowed to the mainline ITER (as discussed earlier, that is, in my opinion, a big mistake, because even if ITER is a successful experiment, the size and cost of an ITER-type power plant makes it unattractive for commercial power). Attempts to move to ammonia borane to gain DOE favor for our all-liquid fuel cell research are underway, but we have yet to obtain DOE support. Simultaneously, Nie and I have moved towards a modified "flow cell battery" type of unit to use with intermittent wind and solar generators. Such storage technology has the potential of gaining importance as more emphasis is placed on developing renewable energy in the United States and elsewhere.

One distinguishing characteristic of our fuel cell research is that we have tackled development of complete systems rather than just one or two of their components. Many other university researchers tend to specialize in fuel cell catalyst development, membrane development, etc. In contrast, our fuel cell research involves developing all of these aspects simultaneously and integrating them into a full system. That can be a daunting challenge. While some work had previously been done on an all-liquid borohydride fuel cell elsewhere, it had not been too successful. One reason was that the catalysis employed was a carryover from traditional fuel cells and not optimized for this unique type of cell. Thus, when we started this research, the first task was to develop new catalysts that would react with the fuels to produce hydrogen and oxygen ions that could directly pass through the membrane without evolving gases. This characteristic is the key feature of the "direct" borohydride/peroxide fuel cell versus one using gases that must be catalytically converted to ions to transport through the cell membrane. Some others such as the Millennium Cell Corporation (prior to going bankrupt) use borohydride as chemical storage for hydrogen, evolving gas from it "on demand" to feed into a standard hydrogen/oxygen fuel cell. If platinum, the favorite catalysis for many processes, is used, gas evolution is vigorous and cannot be avoided as we want to do in the "direct" type borohydride fuel cell. Nie and I, along with Joe Mather, a graduate student from Mechanical Engineering, rapidly screened a number of alternate catalysts in search of one that would efficiently provide direct ion production without gas evolution. Such a study seemed formidable and might take years. However, we combined physical insight, intuition, and trial and error techniques to quickly come up with a set of palladium and gold alloy catalysts for the two liquid fuels involved. This was accomplished in the first eight months of the DARPA project, allowing us to compete successfully for a Phase II project funding. The rest is history.

Digressing for a moment, Joe Mather's contribution to our work, as just noted, has been significant. I first met Joe when he unexpectedly showed up at my office in response to an advertisement I placed on a bulletin board in the Mechanical Engineering building seeking students for participation in fuel cell research. The name "Mather" immediately caught my attention — Joe Mather was the well-known inventor of the "Mather-type" plasma focus developed at LANL. I brought this up with Joe and found that the inventor was his grandfather! I immediately jumped to the conclusion that Joe wanted to join my fusion plasma research. Wrong! He wanted to stay clear from that area to avoid being in his grandfather's "shadow." It turned out that Joe was in the control systems specialty in Mechanical Engineering and he proposed work on controls for our fuel cell. I explained we were not there yet, but had other interesting problems that were associated with Mechanical Engineering aspects of the fuel cell. After some weeks of thinking this over,

Joe joined my fuel cell work. At that moment, Nie Luo and I were focused on finding appropriate catalysts. To our surprise, Joe took an interest in this and quickly reviewed background articles to come up to speed. He ended up doing much work to find an optimized catalyst for this unique all-liquid cell. His work was instrumental to our cause and ended up as his MS thesis. Joe's wife was in medical school and just as he finished his MS she obtained a position in St. Louis. Thus Joe moved there and I lost track of him. However this is another example of how someone with sound basic education who also has great enthusiasm can make great advances in the field.

We tackled other aspects of the cell development in a similar fashion. For example, we needed to go from a 50-W single cell to a 1-kW stack to satisfy contract requirements. That seemed simple at first. Conceptually one could just stack cells together electrically (as done with batteries in flashlights) to increase the output voltage and current. However, for compactness, the flow channels for delivering the fuel to the catalysis layer must also be integrated into the design of the flow plates separating the individual cells of the stack. Plus, a number of mechanical issues are involved related to seals between cells, control of the fuel flow through the diffusion layer, prevention of oxidation of the metals used for electrical contacts, minimizing the pressure drop in the channel pattern to keep the pumping power reasonable, etc. Recognizing these challenges associated with the building of a stack, we decided to visit the Gas Research Institute Lab (GRI) in Chicago. GRI staff had considerable experience with stack designs and their testing. Plus we thought they might collaborate with us in future projects. During the visit and after some discussion, the fuel cell section leader at the Institute told us bluntly that we were "overly optimistic" in thinking we could move to a successful stack in the short period of time we had allowed in the project schedule. (DARPA is known for pushing its contractors to keep exceedingly tight time schedules, and that strategy has allowed DARPA to achieve some important successes. Our fuel cell project was treated in that fashion, with tight schedules and very demanding goals.) Our host at GRI emphasized that there were many non-obvious technicalities in constructing a successful stack that strained the ability of fuel cell experts, especially with a novel new design such as ours. This was a sobering thought, but on the drive back home we all decided these concerns would not deter us. Rather this warning made us even more determined to be successful with the stack development.

Indeed, as warned at GRI, we did encounter several unanticipated problems in the stack design. The most prominent involved the high electrical conductivity of the liquid fuels that allowed electrical conduction through the flow channels that "shorted-out" the cells electrically. We should have realized this in advance, but discovered it quickly in a first test run. We overcame the problem with the "quick fix" of connecting the flow between cells though long tubes to increase the electrical resistance by the long liquid flow path. These tubes came out of each flow channel between cells, went in a half circle, and then led back into the channel of the neighboring cell. The many tubes running in and out were not an eye-pleasing design (fondly termed by our group "the occupus"), but that allowed us to get test data without redesigning the stack flow plates. Later stack designs, like the one shown in the beginning of this chapter, have lengthened flow paths cut into the flow plates themselves, eliminating the need for external tubing. Then, after fixing some seal-related problems, we performed a successful test run at a kW in "record" development time. The fuel cell group held a celebration that night!

To put our fuel cell work in perspective, as already noted that there is considerable work on fuel cells at the University of Illinois, but it is largely focused on basic studies of components. In addition, two totally new total cell designs had been developed prior to our work. They are the direct formic acid cell and the micro laminar flow "membrane-less" type cell. Thus, we are proud that our new all-liquid fuel cell is the third new type fuel cell developed at the University of Illinois.

Several years ago I initiated a class on the "Hydrogen Economy and Fuel Cells" as an effort to bring more nuclear students from NPRE into the area. This course fits into NPRE better than one might think by the title. The point is that nuclear reactors (both fission and fusion) offer an environmentally attractive path to large scale hydrogen production for use in fuel cells. This route would use heat from the reactor to run a high temperature water electrolysis unit. Alternately some researchers are working on a chemical process where the chemicals involved serve as a catalyst to allow water dissociation efficiently at moderate temperatures. Due to its breadth, the course attracted students from other departments as well as NPRE. The hydrogen economy and fuel cell course has now been formalized as a permanent course offering, but the curriculum committee asked that the word "economy" be removed from the name. This is common terminology for people in the community, but the committee felt that the terminology might be confused with financial "economics" as opposed to its meaning of a global energy system based on hydrogen as the prime medium for distribution and storage of energy.

As I worked more in the fuel cell area, I soon realized that another major need for a future self-sustainable energy society is efficient large-scale energy storage technology. This becomes particularly crucial as we push to have "green" energy sources like solar and wind play a larger role in the energy mix in the United States and elsewhere. Management of a power grid where such sources supply a large fraction of the power source becomes unrealistic with their inherent intermittent operation. Adding more base load power plants on the grid to stabilize it can alleviate the problem, but raises the overall cost of electricity due to the excess investment involved in plants that have significant idle time. Some attempts at energy storage have employed truckloads of lead acid batteries or sometimes lithium ion batteries. These "temporary" solutions are unwieldy and expensive. Some locations have natural features that allow traditional methods like pumped water storage using a reservoir, but not many sites have this luxury. Thus research is accelerating to find better solutions employing a wide variety of methods based on chemical, electromagnetic, mechanical, and other technology. Like energy sources, energy storage may end up relying on combining a variety of methods into a grid. The design and methods to control a variety of resources in a grid has gained the name "Smart Grid." In view of the importance and growing research and development on energy storage, I proposed a new course on "Energy Storage" at the University of Illinois. It has been taught several times as a special topics course and should become a formal course soon. As Emeritus, I have played a role in its planning and teaching, but Professor Magdi Ragheb has taken the prime responsibility for it. Among other things, he has posted detailed lecture notes on the course web page.

Because there are no books that fit this course well (several on energy storage tend to focus on a specialized aspect of the problem), the course web page has been instrumental to the success of the class. Professor Ragheb has spent much time developing these extensive lecture notes for the web, and the plan is to eventually convert them to a textbook.

An interesting event occurred in the middle of my fuel cell research. I received a phone call from a venture capital (VC) firm in Dallas, TX, asking if I would consult on the evaluation of a project they were considering that involved "Brown's gas." I knew a little about Brown's gas — basically this gas is the equi-molar mixture of $H_2$ and $O_2$ obtained from electrolysis of water where the off-gases were not separated but collected together. However the original inventor of the gas, Brown (and researchers who later worked on this) claimed that when done in an a.c. electrolysis mode using carefully designed electrodes, the off gas contained a high energy content that upon recombination results in unique bound states of the hydrogen and oxygen. Thus some researchers, including the CEO of the company proposing this project to the VC talked about obtaining H-O-H molecules which were structurally different from normal $H_2O$. A study done in the chemistry department at the University of Miami was cited as backing up this claim. I was fascinated and agreed to do an independent evaluation.

The VC firm then arranged to have an electrolysis unit used for Brown's gas shipped to me in Illinois. We were allotted 3 weeks for the evaluation so it had to be done quickly. With the assistance of workers in my fuel cell group we did three things: added instrumentation to better measure the a.c. power input; used Nuclear Magnetic Resonance (NMR) to search for H-O-H in the off gas; and connected a welding torch to the off gases to evaluate their use for cutting and spot welding of metals. The first two measurements were used in energy balance calculations to further evaluate claims about the benefits if the off gas was used in internal combustion engines for automobiles. I concluded that the use offered little advantage when the complete input of electrical power to create the Brown's gas was considered. Thus I recommended against this application even though it was technically possible to use the gas in an internal combustion engine. I found the second proposed application in a welding torch much more interesting. We performed cutting and spot welding operations on a variety of metals and compared the results using detailed photography and scanning electron microscope analysis versus similar operations done with a standard acetylene torch. Cuts with Brown's gas were much cleaner and sharper than with acetylene. The Brown's gas flame exhibited clean cuts with very sharp edges of the metal instead of a melted ridge. Spot welded samples were also compared using destructive testing and found to be superior. The reason is tied to the mechanism for transporting heat from a Brown's gas flame to a metallic surface. This occurs via rapid recombination of the hydrogen and oxygen to release energy while forming water upon contact with the surface. Indeed, it had been reported that one could pass their hand through the Brown's gas flame without feeling the heat. I personally verified this, but only once! The explanation for this unexpected feature is that rapid recombination of the gases does not occur on one's skin like it does on metal which serves as a catalysis for recombination. Obviously rings should be removed from one's fingers before attempting such tests! Also, severe burning can occur if one mistakenly uses an acetylene torch.

I reported to the VC that this application offered promise. The VC then arranged to have a Brown's gas torch sent to a large metalworking facility near the Dallas-Fort Worth airport. Managers of this facility agreed to have several of their welders try it out. I flew down to observe the tests and to be involved in discussions with the facility managers. The tests went well, performing much as I previously found. However after considerable discussion with the facility management, they concluded that the improved cutting and welding features were not so overwhelming to warrant changing from acetylene (or in some cases from laser cutting). Their workers were used to using acetylene, and it was easy to obtain acetylene tanks and transport them to work sites. In contrast, the Brown's gas electrolytic unit required electrical connections and had to be turned on some time in advance before delivering gas to the torch. While not major obstacles for use, these factors were considered to be disadvantages, and so the managers finally concluded that they might only use one or two units for special operations. Thus the VC ultimately decided not to invest in the company proposing this progress.

I should, however, mention another aspect of this experience. Gabriella Drancy, a talented young lady employed by the VC company in Dallas to handle field negotiation for these evaluations and some other proposed projects, came to all of the tests. During discussions with me she also learned about our fuel cell work and expressed interest in it. Her plan was to return to school in the fall at Southern Methodist University to earn a MBA degree. As part of her classes she was required to prepare a business plan for a start-up company. Gabriella proposed that she use our fuel cell as the basis for her business plan. I was pleased to have her do that because having such a plan would be a big help as we moved NPL Associates, Inc., in that direction. We had many email discussions to provide information for the plan. Gabriella also attended a fuel cell conference and exhibition with us to learn firsthand more about companies in the field. This plan turned out to be very useful for NPL and also resulted in an A grade for her! It is interesting how so many different directions for further study sprung from this short evaluation. Many open basic questions about Brown's gas and its production remain, but I had to return to my fuel cell research.

## Further Reading

G. H. Miley, N. Luo, P. J. Shrestha, R. Gimlin, R. Burton, J. Rusek, and F. Holcomb, "$H_2O_2$ Based Fuel Cells for Space Power Systems," *AIAA International Energy Conversion*, San Francisco, August (2004).

G. H. Miley, "Optimization of Catalyst Deposited Diffusion Layer for a Direct Sodium Borohydride Fuel Cell (DNBFC)," *ECS Transactions — Fuel Cell Seminar & Exposition*, Vol. 12 (2008).

G. H. Miley, N. Luo, and K. J. Kim, "A Dry-Borohydride/Injected-Hydrogen-Peroxide Fuel Cell," *ASME Conference Proceedings, FuelCell2010*, 33259, pp. 119–125 (2010).

G. H. Miley, M. Ragheb, and N. Luo, "Experience Teaching an Energy Storage Class as a Nuclear Engineering Course," *Trans. of the ANS*, Vol. 105, pp. 125–126, Washington, D.C., Oct. 30–Nov. 3 (2011).

# Chapter 14

# Fusion Propulsion and Space Colonization

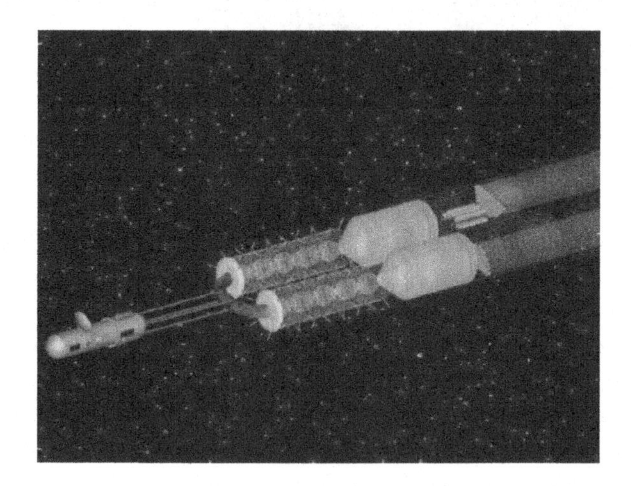

*An artist's image of Fusion Ship II, a 750-MWe IEC fusion powered spacecraft with ion thruster propulsion intended for manned space missions. Ten coupled IEC fusion units allow for deep space missions (the ball-like structures to left of funnel-shaped magnetic nozzles where the exhaust plasma flow produces the thrust needed for very fast travel). This design was worked out in considerable detail in collaboration with Hiromu Momota (a visiting Professor from Japan) and Rod Burton (a Professor from our Aerospace Engineering Department) plus students from the IEC group. It represents one of the highest performance designs studied for manned flight with fusion propulsion to date. Thus this research gained considerable attention in the advanced space propulsion community and has been presented at invited talks at several major AIAA and ANS meetings.*

Anyone working on fusion research sooner or later becomes interested in fusion propulsion. It is widely recognized that fusion-powered space propulsion is one of the "best" choices for deep space travel, combining the advantage of high specific impulse (i.e., a high exhaust velocity) to enable speeds approaching the speed of light along with the high power needed for takeoff and landing. The famous Daedalus concept, published in an early report by the British Space Society, employed a "miniature hydrogen bomb" explosion technique to exploit this capability with "existing technology." This scheme used repetitive bomb explosions behind the space ship, supplying propulsion by directing debris against a "pusher plate." Later studies by others moved away from the use of bombs

*I served on the Air Force Research Advisory Board for about ten years. This board was charged with advising the Air Force about research planning, both for the research labs at base locations and for the Air Force Office of Scientific Research which funds university and industrial research contracts nationwide. Members included high ranking officers and also civilians from Air Force labs and universities. I took this photo of attendees during a meeting in Washington, DC, January 17, 1991. These meetings traditionally ended with a dinner for the board and associated staff. One of these turned out to be especially memorable when a Lt. Colonel came running in to tell a four-star General eating with us that CNN TV had just announced that the United States had bombed Iraq, starting Desert Storm (a strange source for this news considering the extensive intelligence network of the Air Force). The dinner was hastily concluded, but as we left the building (in the center of various state department and defense department buildings in downtown Washington DC) we found that security forces had already closed off the streets in front of the building. The whole area was flooded with high intensity lights as a security measure for the important government buildings next door. To exit, we had to undergo a complete security check. Nothing happened that night, but apparently there had been some concern that terrorists might try to respond to the bombing by sabotaging one of these vital government buildings.*

to consider conceptual designs for fusion reactors to power such rockets. Thus, when I entered fusion research, I read these reports and became fascinated with this potential future application of fusion. However, I was first introduced to nuclear propulsion in a serious way earlier when I was interviewing for jobs as I neared graduation with a PhD. When I went to LLNL on an interview trip, the nuclear fission (not fusion) rocket group had an opening that they discussed with me. Indeed, had I gone to LLNL then, I might have ended up in that project. However, as noted earlier, I finally accepted a position at KAPL.

LLNL was working with LANL on fission reactor propulsion under the Nuclear Engine for Rocket Vehicle Application (NERVA) and Project Rover programs. R. W. Bussard (often referred to by friends simply as "R. W.") was then head of the LANL NERVA project. They were building fission rockets that used a fission reactor to heat hydrogen flowing through it, which was then exhausted from a magnetic nozzle for thrust. These rocket units were mounted on flat bed train cars, which were tested on railroad tracks placed across the Southwest desert. The tests seemed quite successful, but some of the fuel rods overheated and released small amounts of radioactivity that spread out across the desert. In those days, no one was around to be affected or to complain, and this was not viewed with the alarm it would receive today. Ultimately, R.W. Bussard along with his co-worker, R.D. DeLauer, wrote a pioneering book titled *Nuclear Rocket Propulsion*, published in 1958 by McGraw Hill. I bought a copy after the LLNL interview and still have it today. Later Bussard became famous for the "Bussard Ram Jet" concept for interstellar flight which incorporated a design that "scooped" up hydrogen in flight and compressed it to fusion temperatures. The Bussard Ram Jet was featured in science fiction stories and films at the time. I did not realize then that some years later I would meet and become a collaborator with "R.W." when he became heavily involved in fusion. His initial work in fusion was in the DOE Office of Fusion Energy, where he headed the Fusion Technology Office. Then he set up a privately funded "disposable" high field Tokamak project, and later founded the company EMC$^2$ to do IEC (Polywell type) fusion research as described earlier in the IEC Chapter. I held a DOE contract from his office and later collaborated on IEC research when he formed the company EMC$^2$. Indeed when R.W. discovered he was dying from cancer he asked me and several others for advice about finding someone to take over technical leadership of EMC$^2$. I recommended my former student Rick Nebel, who was doing IEC work at LANL. Rick ended up taking over EMC$^2$ while R.W.'s wife, Dolly, remained the owner and handled financial matters.

I would note that fission propulsion is exciting for near-term application to missions like returning to the moon and possibly for further out trips to places like Mars. It offers higher exhaust velocities than a chemical rocket, has a high power capability, and was demonstrated in principle in the early Rover/NERVA tests. However, fusion rockets offer the much higher velocities needed for deep space travel. This is because the fusion system can exhaust the hot plasma and energetic fusion products out through a nozzle directly without heating an intermediate propellant gas flowing past the hot fuel elements as done in a fission rocket. In the latter case, the maximum temperature is limited to prevent melting of the fuel rods, and that in turn limits the exhaust velocities. There are some fission rocket concepts like the gaseous core reactor or the fission product plate design that might get around this limitation, allowing exhaust velocities near to those of fusion. However, these have not been studied to any extent experimentally, and have serious development issues. Thus in my opinion it would be better to put effort into fusion systems than start such programs — but the basic fuel rod fission design still could parlay nuclear propulsion into an intermediate role while the longer term fusion development is under way.

My interest in fusion propulsion was again rekindled as I undertook Mirror Confinement and Field Reversed Configuration (FRC) fusion research. Both offer natural magnetic nozzle "thrusters" through their "built-in" diverter configurations. LLNL workers had already done some design studies of mirror propulsion units which looked promising. I decided to undertake a FRC design study, because its combination of closed plus open magnetic field structure potentially offered a higher energy gain than a standard mirror along with a natural diverter action for creating thrust. Shortly after doing some initial work on this, I visited Professor Hiromu Momota at the plasma physics laboratory at Nagoya University in Japan where FRC research was becoming a major topic. He, in turn, introduced me to researchers from other Japanese universities collaborating with him on FRCs.

That led to a number of important contacts and allowed me to set up some additional collaborations on FRC research. Later, Hiromu and I visited one of these groups at Kyushu University. During the visit I met Hideki Nakashima, then Assistant Professor in the Department of Energy Conversion Engineering. He was extremely interested in fusion propulsion using the FRC. Subsequently, Hideki and his wife came to the University of Illinois for a year to work with me on FRC propulsion. This resulted in an updated conceptual design, which highlighted both the promise of this approach and the various physics and technical hurdles to be overcome before a power unit can be developed. One of the key unanswered issues for the FRC is its stability in high power level units. The configuration is naturally MHD unstable, relying on inertial effects from orbiting ions for stability. Indeed, as noted earlier, Ed Morse tackled this problem as part of his PhD thesis and his computations provided important new insights into windows of stability. Since then, many other stability studies have been done and much more understanding has been gained. Some major advances were instigated by the ARTEMIS D–$^3$He FRC reactor study by Hiromu Momota and Japanese colleagues discussed earlier. I attended several meetings to discuss the status of FRC stability with Hiromu in Japan during that project. Loren Steinhauer, of the University of Washington, and Herb Berk, of the University of Texas, also came from the United States to attend these meetings. Both of these well known plasma theorists went on to make important additions to FRC stability theory. Experiments have also provided important insights. Some experimentalists have produced FRCs (using non-fusing gases) and translated them from one chamber into another. Also, small FRCs have been injected into Tokamaks to provide combined fueling and helicity (twisted magnetic field) injection. Much of that work has been pioneered by the fusion lab at the University of Washington.

Interest in FRCs for various applications continues today. On a visit to the Edwards Air Force Base (EAFB) in November 2009, I consulted with workers doing an experiment exploring electrically driven FRCs intended for satellite thrusters. Still, despite all of this work over the years in various labs, DOE has never found the money to explore this very interesting "alternate confinement" concept (or other so-called "innovative" confinement concepts) in a reasonably-sized experiment. Thus, it and related concepts are viewed with skepticism by DOE managers and also by Tokamak "fans" in the DOE labs. The attitude is that plasma problems (stability, etc.) are bound to occur, as happened in toroidal research as one scales up in size and power level. This poses a catch-22 situation. Without the money to try, the issue cannot be resolved. These circumstances have remained for years in the magnetic fusion programs as all major efforts have concentrated on Tokamaks. ITER, for instance (which is not even a true "demo" reactor), costs three-hundred billion dollars and will not be ready for plasma tests for about 20 years. Consequently, as previously mentioned, some years ago Larry Lidsky wrote an article asserting that the ITER is a "white elephant" (i.e., too large, complicated, and expensive to be used by energy companies).

Another fascinating topic associated with space fusion power plants is D–$^3$He fuel and the possibility of mining the moon for $^3$He. Indeed, I have been quite enthusiastic about D–$^3$He for fusion since, next to D–T, it enjoys one of the highest fusion cross sections. However, $^3$He is not available beyond laboratory scale quantities on earth. Thus, I had proposed to develop a series of $^3$He breeder plants burning semi-catalyzed D–D fuel to supply $^3$He. Critics objected, however, saying that this required development of two different types of fusion plants — the breeder and the D–$^3$He burner. In characteristic fashion, I argued that this would not be so difficult due to the overlap in technologies of the two. Thus, I was amazed and delighted when Lloyd Wittenberg and colleagues from the University of Wisconsin submitted a paper to *Fusion Technology* that I edited about the exciting possibility of obtaining $^3$He by mining the upper sand/gravel crust on certain regions of the moon. $^3$He has been bombarding the moon's surface for ages, coming from fusion reactions in the sun and carried by the solar "wind" to the lunar surface. Particles in this wind easily reach the lunar surface due to the lack of a protective layer of atmosphere such as we have here on earth. Wittenberg and his colleagues made a very convincing case that the mining and handling of this $^3$He could be done economically, including transport back to earth. The amount available, in terms of fusion energy content is amazing, far more than the energy content of earth's oil reserves. As previously mentioned, Jack Schmidt, a geologist-turned-astronaut, had taken samples from the lunar surface during Apollo 17 that led to this discovery. Analysis of these samples showed this soil contained, among other gases, $^3$He. At the time, however, the significance of this was not recognized. Later, Schmidt became an Adjunct Professor in Geology at the University of Wisconsin where interactions with their strong fusion program brought all of this into stark realization. In recent years Schmidt has given a number of impressive talks around the nation explaining the importance, in his mind, of returning to the moon to mine $^3$He. He also did a short stint in Congress as a representative from New Mexico and tried to spread the word there. However, due to its long-term nature, the issue never caught fire in Congress.

Lunar mining of $^3$He remains somewhat controversial, though. Some say this is the most valuable resource on the moon, thus it could justify future lunar missions. Others say it is of little use because we do not yet (and may not for years) have a fusion reactor to burn it. Others worry that the "strip mining" needed to mine $^3$He would seriously damage the environment of the moon, including formation of an atmosphere as many other gases in the crust are released in the process. Yet, others simply refuse to accept the economic studies and think $^3$He would be too costly to use for commercial power. I, personally, come down on the positive side of mining. However, I believe that with the slowdown in the space program, mining is yet well into the future. Thus, I have returned my focus in this area to the $^3$He breeder concept. I view it to be a faster route to building D-$^3$He fusion power plants for both terrestrial and space use.

As a member of the AIAA Colonization Technical Committee, I learned much from others about the opportunities and challenges of establishing colonies on planets like Mars. I took special interest in power sources for these future colonies. Several of my colleagues and I recently completed a chapter in a handbook published by Springer Science about the colonization of Mars. In that chapter, we speculate about a system of larger p–$^{11}$B IEC fusion plants that would supply electricity and also be used as a plasma source for various base manufacturing operations. The first IEC would be shipped up from earth, but thereafter, the IEC would provide electrical power and plasma processing to use indigenous materials on Mars to manufacture the next units. These plants would be supplemented by use of small LENR "battery-like" power units for mobile use. While only a vision at this time, such visions have great value to enthuse and provide a roadmap for future developments. I was greatly inspired in this effort by my two co-authors, Xiaoling Yang and Eric Rice. Xiaoling, then a post-doc, and Eric, CEO of the ORBITEC company in Madison, WI, became very enthusiastic about the concept of using these power units for colonization. I might have backed out of the project without their enthusiasm. Eric Rice had a strong influence on me for same years, not only in the article but in encouraging me to do space studies and to include colonization issues as an interest. Eric, as previously mentioned, is a former University of Wisconsin football player who "tackled" the business world, becoming the CEO of ORBITEC, a very successful small aerospace company. He has been very active in AIAA activities and upon founding the AIAA Mars Colonization Technical Committee (TC) got me involved. (AIAA TCs are formed by group members in areas of their interests. TCs provide position papers on select subjects and arrange sessions at various AIAA meetings.) Others too have learned it is hard to say "no" to Eric once he has decided to do something with your help! While working on the handbook article, Eric became interested in LENR reactions and is pushing to undertake a joint project with me on that. At any rate, fusion and space seem to be a natural fit (after all, fusion is nature's way in space). Time will tell how man-made fusion works out there. Hopefully, my visions will add to those of many other space enthusiasts in their effort to turn a dream into reality.

I still continue fusion space propulsion research to this day. These studies have evolved through a variety of fusion confinements systems to use as the propulsion unit. They have included conventional mirrors, Field Reversed Mirrors (FRMs), Spheromaks, the Dense Plasma Focus (DPF) and various forms of IECs. These design studies were intended to be "path finding" for the respective concepts, i.e., to identify the issues and direction for a development program for them. In our design studies we have usually focused on key plasma physics issues that are, to some extent, generic for all the concepts. The confinement concept employed in the design study has often been one that we were also doing theoretical and/or experimental work on. This provides the designs with a reasonable physics base. An overriding theme has been the need to utilize advanced fuels like D–$^{3}$He and p–$^{11}$B to reduce complications of neutron fluxes, neutron activation of materials, and tritium management. In addition we have focused on using the energy carried by the charged-particle reaction products for direct thrust and also for direct conversion into auxiliary electricity. The lessons learned from the studies are many and too involved to discuss here. Next I will comment on a few events and people that strongly influenced my work and views on fusion propulsion.

Indeed, fusion propulsion and use of D–$^{3}$He are not new ideas. In the 1970s, the Director of the NASA Research Center near Cleveland (the Lewis Research Center, which is now the NASA Glenn Research Center) called me and proposed to hire me as a consultant to review their fusion propulsion program. It was under attack by a NASA committee charged with selecting projects to cut to save money during a downturn in their budget. The NASA Glenn management hoped that my objective evaluation would turn out to be positive and "save" the program. I accepted the task and spent some time reviewing their work. The project was very impressive, with the highest large-bore superconducting magnet array built anywhere up to that time. The confinement scheme used multiple mirror confinement units linked together in a toroidal configuration so that the plasma leaking from one entered into the next one. This is called a "Bumpy Torus" and is intended to reduce the excessive leakage from a conventional single mirror. Such a configuration is unstable, however. The Lewis device used electrostatic field stabilization with a high voltage ring-like electrode embedded in the throat end of each mirror. The staff, headed by Jack Rienmann, included a bright, enthusiastic young plasma physicist, Reese Roth. Reese was not only a fusion scientist, but also energetic proponent of fusion space propulsion. Even in those early days, Rienmann was interested in using D–$^3$He as the fuel and had several of his staff exploring that option, although all experiments used a hydrogen plasma to avoid the expense of $^3$He and to avoid the need for neutron shielding. Measurements of densities and temperatures in a hydrogen plasma provide valuable data about the plasma heating and confinement. Thus most fusion experiments use this approach. I was quite impressed with their accomplishments with this experiment. However, one "political" problem became apparent. The group was somewhat isolated from the larger outside fusion community, partly because they enjoyed a reasonably constant (until then) NASA funding and did not get involved to any extent with other fusion labs. Most of the outside community was funded through DOE and many fusion scientists were at DOE fusion labs. As a consequence, the DOE personnel were in good contact with each other through various DOE meetings that did not include the NASA group. This was not necessarily done on purpose but was a natural consequence of the separate funding agencies. Still, this left the NASA group as "outsiders."

To further complicate matters, Oak Ridge National Laboratory (ORNL) had a competing experiment called the "ELMO Bumpy Torus" headed by Lee Berry. This device used an RF induced electron ring for stabilization, thus avoiding the solid electrode structure of NASA's experiment. At the time of my review, the ORNL group was claiming great success and planned to request money from DOE to build a larger device. Plus they argued that internal electrodes used in the NASA Bumpy Torus could not survive in a fusion plasma (the NASA group disputed this due to a magnetic protection system they devised). A majority of the fusion community knew the ORNL people better and sided with them in this somewhat heated debate. That did not help the case for the embattled NASA group with the budget review committee. After my detailed review I ended up writing a quite positive report for NASA management, but to no avail — the project was canceled with others in a round of NASA budget cuts of advanced research and development projects. Ironically, some months later, the ORNL group publically disclosed that they had found an error in their diagnostics which led them to report erroneously high electron temperatures. This disclosure ruined their case for a new larger experiment, and eventually the ELMO Bumpy Torus was shut down. This is one of the few times in my years of fusion research (or in other research areas) that I have seen an open admission of an error. As discussed later relative to my oversight duties on the safety panel (called NRAG) at the nearby Clinton fission reactor power plant, I have found that it is human nature to rationalize such things to avoid receiving blame. My job on the NRAG committee was to try to cut through this reluctance and get to the "root cause" of events. I don't know exactly what happened in the ELMO project, but the honesty of the ORNL group is certainly commendable — if done more frequently, tax payers would have saved much money!

As an aside, I would note that Reese Roth, who worked on the NASA Bumpy Torus project, moved to the University of Tennessee as a Professor of Electrical Engineering. There he diversified his research to include plasma processing as well as fusion plasma physics. Thus I often saw Reese at various technical conferences and we have remained friends.

In a very unfortunate event that occurred several years ago, Reese was fined and given a jail sentence for a violation of an International Traffic in Arms Regulations (ITAR) restriction on an Air Force contract he held. Presumably he had a Chinese student on the project and also gave a seminar in China regarding the work — both of which were viewed as violations. Many scientists who knew him, including myself, wrote letters to the judge requesting that he be lenient in view of Reese's long years of research service to the United States. In my letter I added that I was sure that Reese would not knowingly violate these regulations. Rather, he must not have understood the restrictions put on the contract. However, the judge apparently wanted to make this case a warning to others, handing down a harsh sentence of several years in jail plus a fine. Reese began his jail sentence in February 2012 after his appeal was rejected. The situation makes little sense to me in comparison to cases reported in the news media where large companies have simply received modest fines for much more serious violations, including actual hardware transfers. The U.S. attorney who prosecuted the case is quoted as saying that this was the first use of ITAR to prosecute sharing classified data with foreign citizens. (This is a problem in many universities. The University Grants Office usually points out the ITAR restrictions to the principle investigator in advance. ITAR restricted projects are difficult to handle in an university environment where many international students are enrolled. Indeed, sometimes at Illinois the restrictions cause the Grants Office to refuse the contract!)

At one point, as described in Chapter 8, I undertook a fairly large effort in dense plasma focus (DPF) research with support from Frank Mead from Edwards Air Force Base. I mention that again because this research was largely motivated by the potential for use of the dense plasma focus for space propulsion. Like the other alternate confinement concepts discussed here, the DPF is characterized by not requiring heavy magnets or chamber structures and by the ability to burn advanced fusion fuels. These features make it another good candidate. Funding problems finally stopped my work. However, following a brief collaboration on the project, Eric Learner started a private company and nonprofit foundation in New Jersey to support an aggressive plasma focus effort. Our association began when Eric called me while we still had our DPF experiment running and requested rental time on it to do experiments under a NASA SBIR grant he held. Later he spent a month or so at Illinois doing these experiments. We helped him and became friends. Eric and I did not agree on all of the physics of the focus, but we did learn from one another during that period. Eric did not have formal training in science but was self-taught following a brief time as a science fiction author. Anything he lacked in formal training was more than made up for by his vision, hard work, and enthusiasm for space travel.

While I have taught fusion courses at Illinois for many years, there is not one devoted to fusion propulsion. I did manage to slip in one or two lectures on the topic in the advanced fusion course. Also, I had a very interesting opportunity to teach a design class on fission (not fusion) thermal rocket type propulsion in early 2000. Wayne Solomon, then Head of the Aerospace Engineering Department, asked me to join him in teaching a senior-level design class on the subject. Wayne understood the basics of space propulsion well, including orbital mechanics, air breaking strategies, etc., and he had some understanding of fission neutronics. However, he felt that he should have someone with a more extensive reactor physics background involved. I agreed to co-teach the course. We really enjoyed working together and made a great team! It was relatively large for a design class, with about 20 students. The students were well prepared for the aero part of the study and enthusiastic. During that period NASA had announced new interest in nuclear rocket propulsion for a Mars mission. Thus we all felt the class design could have some impact on NASA's plans, making us all more determined to do a pioneering design. Wayne also arranged for several scientists working on the topic at NASA Glenn to serve as consultants to our design class. They made a visit to the class and provided input to help resolve issues encountered as the design progressed. The plan was that they would incorporate some aspects of the class's final report into plans at Glenn. Unfortunately, the last week of class, the news came out that NASA had abandoned plans for nuclear propulsion due to funding cuts. (Seems like an all too often event for these government agencies.) I have encountered this type of "start-stop" government project often, as documented in other chapters of this book. Unfortunately such management of research by our government agencies wastes tax payer money and also causes a loss of skilled personnel in the area should the program be restarted later. In addition to frequent changes in Congress, this type of action results from the lack of a long-term energy policy plus changes in goals for NASA in the United States.

Working with Wayne on this course also affected my space propulsion research. In additional to technical information we learned doing the design study, Wayne introduced me to Rod Burton, Professor of Aerospace Engineering. As described elsewhere, Rod and I have collaborated closely on space research ever since. Wayne also introduced me to David Carroll, who works in the small company CU Aerospace that Wayne founded and setup in the University of Illinois Research Park. David had received his PhD in Aerospace Engineering at Illinois a few years earlier and decided to stay in Champaign because he felt this small company offered a great opportunity. Research underway at CU Aerospace had considerable overlap with my interests. In addition to propulsion research, they were working on an electrical discharge driven iodine laser. This approach is closely related to my earlier research on nuclear pumping of singlet-delta oxygen, as described in Chapter 4 on NPLs. These mutual interests caused continued contact which later resulted in several joint projects. One of these projects involved use of the air independent borohydride/peroxide fuel cell for space applications, described in the previous chapter. Dave is also very active, like I am, in AIAA committees and in the University's Research Park Tech Community Association where we frequently see each other.

While discussing space propulsion, I should mention two other former NASA workers, John Cole and Norm Schulze, who interacted with me over the years. John headed the advanced propulsion effort at NASA Huntsville for many years. He funded a number of projects, including efforts at the other NASA labs like JPL. I received a little funding through him, but as I mentioned earlier, it was mainly our many discussions about advanced space propulsion that influenced me. He had great experience in the area and helped set me straight when I started to go away from the correct direction.

I first became aware of Norm Schulze when he called me in the late 1990s and asked my opinion about future directions and the potential for fusion propulsion. Norm was a director of the operational and engineering department at NASA headquarters in Washington, DC. Advanced propulsion was not part of his formal job, but he had become very interested in fusion, realizing its importance to future NASA deep space missions. Our discussion further reinforced his thinking, so he called me a month later and asked if he could come to Illinois to discuss fusion propulsion in more depth. He felt he lacked a background in fusion but wanted to, as he said, "take a crash course." This quest almost became an obsession. Norm visited a second time, and on that trip continued on to visit several other fusion groups around the country. Later Norm spoke on the topic at various NASA technical meetings and workshops. By then he was completely convinced NASA had to start work on fusion now — as he stated it: "since it (fusion) will take some years to develop one must start work 'now'!" Despite his strong feelings and many attempts, he failed to get the attention of other key NASA managers in the DC headquarters. Thus Norm set about writing a lengthy official NASA report on the subject, explaining the basics and proposing a concrete program plan to start fast-track development. Norm was by then quite convinced the use of lunar $^3$He in this program was essential, so the report also addressed lunar mining. The report, while masterful, still did not change the minds of NASA management. There were two main reasons — most officials remained focused on near-term issues and DOE, not NASA, remained the government agency charged with fusion development. A joint DOE-NASA office to spearhead this effort seemed essential to move ahead. Such joint offices for specific projects have precedence, but the formation of one for fusion development faced a political struggle between personnel in the two agencies. There was not an interest in undertaking that battle, so once more nothing happened at NASA! Norm was disillusioned and eventually retired. We still interact off and on and exchange Christmas cards.

My present plasma propulsion project builds on long term interest in the IEC (see Chapter 11). This experimental program backs off from fusion propulsion and instead is aimed at electric propulsion for next generation satellite positioning. Thus it is in competition with Hall thrusters and related concepts. Our concept, termed Helicon Injected Inertial Plasma Electrostatic Rocket (HIIPER), uses a radio frequency device (Helicon) to generate the plasma that is injected into the IEC. The IEC in turn accelerates the ions in the plasma and forms a very intense small plasma jet which provides the thrust. HIIPER has simple structures, hence is lightweight. It offers good efficiency and a variable exit velocity and thrust power for easy mission control. My strategy behind undertaking this effort is that it offers near-term applications, making funding easier. Yet some of the plasmas physics that will be developed carries over to eventually aid development of a fusion powered IEC thruster. Indeed, we recently presented a preliminary conceptual design of a space probe termed VIPER powered by the fusion version of HIIPER. VIPER demonstrates the tremendous promise for future deep space missions using fusion in an advanced concept such as this. Ben Ulmen, a PhD candidate in NPRE, is doing his thesis on HIIPER. Three other MS candidates, Paul Keutelian, George Chen, and Akshata Krishnamurthy are also working on the project. Much of the equipment involved was obtained from Edwards Air Force Base following Mike

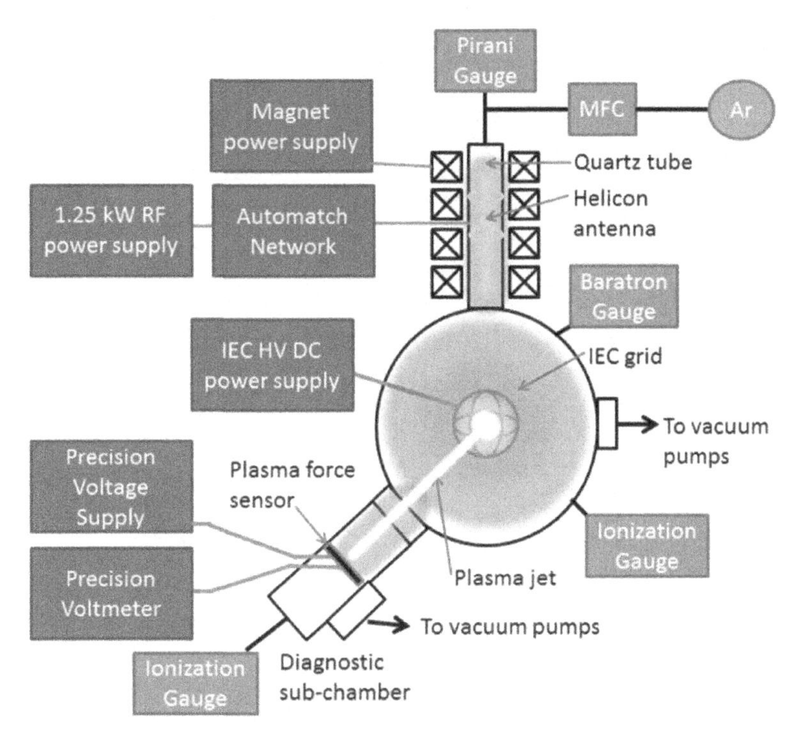


*This is a schematic of the HIIPER. High density plasma is produced by the helicon source and injected into the IEC, which accelerates the ions and forms the jet exhaust for thrust. This jet is in the form of a thin plasma beam produced by an asymmetry in the central cathode grid created by an enlarged grid opening.*

*Left: The helicon plasma source. It has a 13.56 MHz RF power supply that provides 2.2 kW max power with 1 kW max auto-matching network; a quartz tube with copper strap m = +1 helical antenna; and 1200 Gauss maximum electromagnets (water cooled). Right: The IEC device has a custom-built stainless steel grid, shown with the outer chamber open. It has a 50 kV, 50 mA power supply that provides 1 kW max power.*

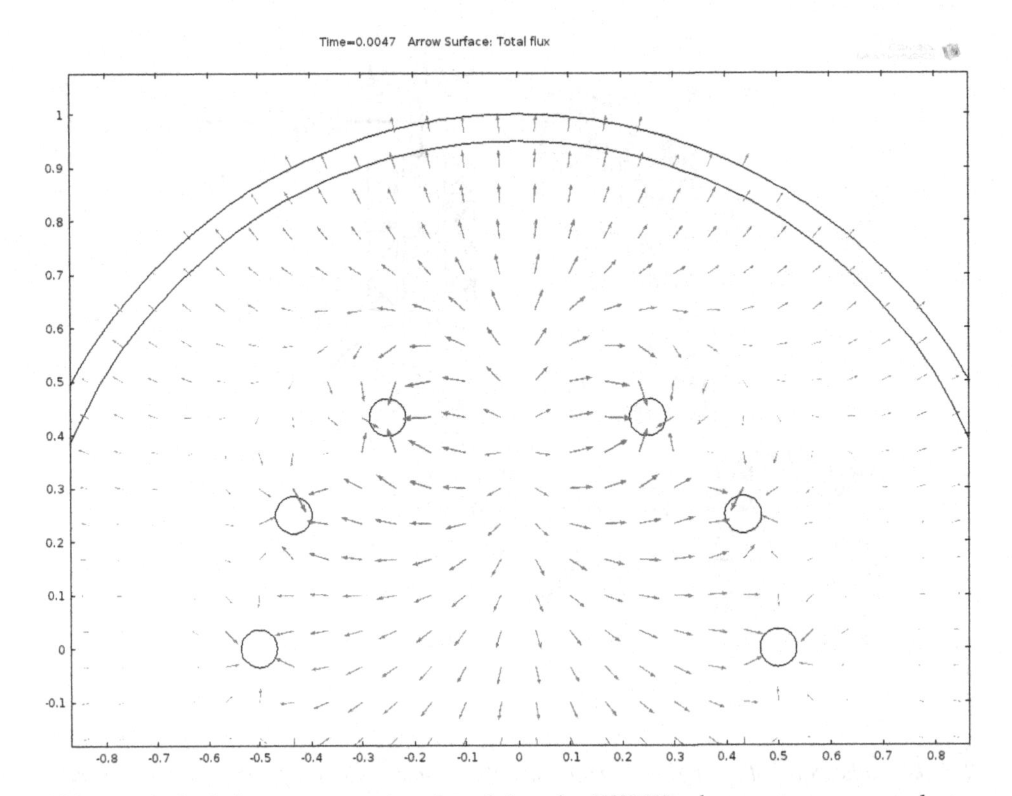


*This graph depicting some results of studying the HIIPER plasma was presented at a recent AIAA propulsion meeting. The arrows in this graph (made using the COMSOL Multiphysics software discussed in Chapter 16) represent plasma flow lines forming the jet.*

Reilly's PhD work there on the helicon plasma generator. Other funding has come from John Scott at NASA Johnson Space Flight Center (see Chapter 9). This has developed a strong interest in aneutronic fusion propulsion and associated direct energy conversion. His encouragement has given a real boost to the project. I am proud of HIIPER, and it may be my last major plasma experiment at the University of Illinois. Thus, to provide more insight into HIIPER, a diagram, photos, and a computation representation of the jet are provided here.

HIIPER is an ambitious program. It and my LENR work represent my main research efforts as this book was being finalized. Thus, my Black Swan search on new energy systems has been narrowed down to two!

## Further Reading

A. Bond *et al.*, "Project Daedalus — The Final Report on the BIS Starship Study," *JBIS Interstellar Studies*, Supplement (1978).

L. Lidsky, "The Trouble With Fusion," *MIT Technology Review,* October (1983).

G. H. Miley, *et al.*, "IEC Thrusters for Space Probe Applications and Propulsion," *AIP Conference Proceedings*, Vol. 1103, pp. 164–174 (2009).

G. H. Miley, J. Orcutt, A. Krishnamurthy, P. Keutelian, B. Ulmen, and G. Chen, "A Fusion Space Probe — Viper, an Ultra-High $I_{SP}$ Pulsed Fusion Rocket," *Space Technology & Applications International Forum (STAIF II)*, Albuquerque, New Mexico, March 13–15 (2012).

G. H. Miley, X. Yang, and E. Rice, "Distributed Power Sources for Mars Colonization," *MARS: Prospective Energy and Material Resources*, edited by V. Badescu, Springer, pp. 213–239 (2009).

G. H. Miley, R. Stubbers, J. Webber, and H. Momota, "Magnetically-Channeled SIEC Array (MCSA) Fusion Device for Interplanetary Missions," *STAIF 2004,* edited by M. S. El-Genk, *AIP Conf. Proc.*, Feb 8–11, No. 699, pp. 399–405 (2004).

N. R. Schulze, G. H. Miley, and J. F. Santarius, "Space Fusion Energy Conversion using a Field Reversed Configuration Reactor," *Penn State Space Transportation Propulsion Technology Symposium*, June 25–29, pp. 453–499 (1990).

B. A. Ulmen, P. Keutelian, G. Chen, A. Krishnamurthy, M. P. Reilly, and G. H. Miley, "Investigation of Plasma Properties in Helicon-Injected Plasma Electrostatic Rocket (HIIPER)," *48th AIAA/ASME/SAE/ASEE Joint Propulsion Conference and Exhibit, 10th Annual International Energy Conversion Engineering Conference*, Atlanta, Georgia (2012).

# Chapter 15

# Nuclear Batteries

Before discussing my interest in work on nuclear batteries, it may help to explain what this technology is by briefly discussing several nuclear battery concepts I have worked on, illustrated in the following figures.

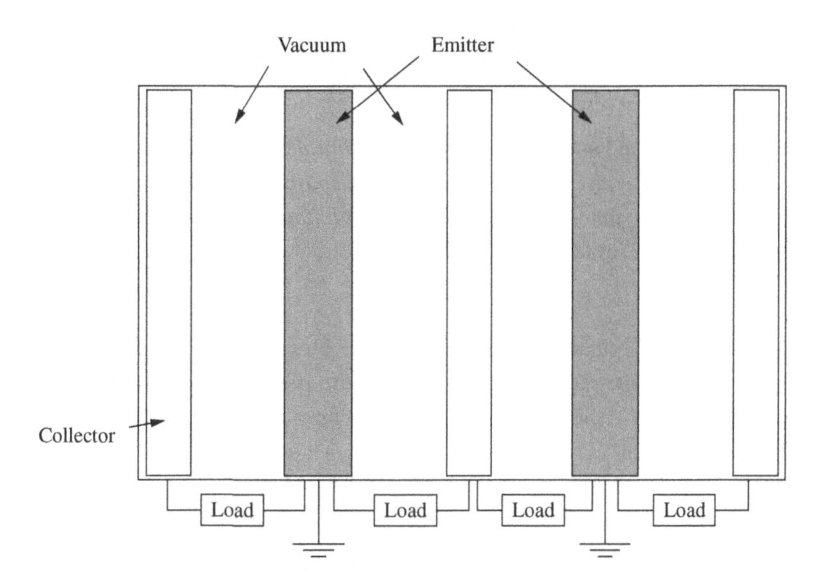


***The Honeycomb Type Nuclear Battery.*** *The emitters are formed by a honeycomb-like structure coated with a Beta-emitting isotope. This allows the emitted Betas to exit on both sides of the emitter, greatly increasing the cell efficiency.*

These figures illustrate two classes of nuclear batteries — direct collection and p-n junction types. Direct collection units such as illustrated in the first figure have a radioisotopic layer emitting Beta particles (energetic electrons) which pass through a vacuum (electrical insulator) to reach a "collector" plate held at high voltage corresponding to the Beta particle energy. Thus the kinetic energy of the Beta particle is converted to high (multi-kV to MV) voltage, giving a dc electrical current output. This battery then is characterized by a high voltage, low current output and quite high conversion efficiency. The design shown in the figure is a unique "bi-directional" direct collection cell. It uses a special honeycomb emitter structure to support a thin isotope layer such that Beta particles can easily exit both sides of the electrode to reach the collector plates. This reduces source losses by almost a factor of two, giving the cell a very high conversion efficiency. The design shown here was part of Jin Lee's thesis work.

However, the very high voltage developed in such direct collection concepts is not desired in some applications. Voltage step-down converters can be used, but they add complexity to the design. Hence the alternate p-n junction type cell shown below has a lower efficiency but offers larger currents at low voltage. Thus it fits more easily into conventional circuits.

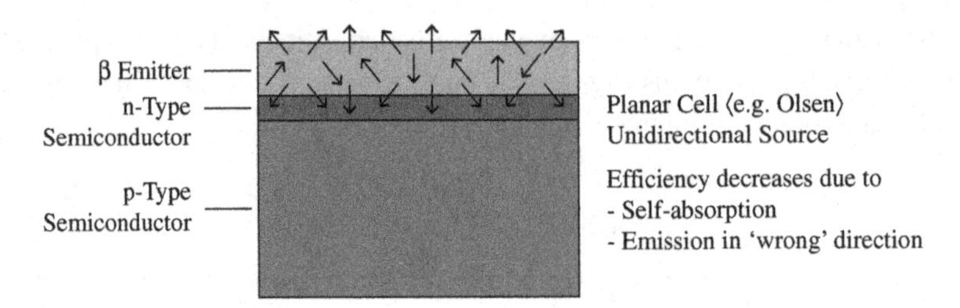


***P-N Junction Type Beta-Voltaic Nuclear Battery.*** *In this cell, Beta particles enter the p-n junction layer producing electricity as is done in a solar panel. However the Betas penetrate much deeper in the junction material than do photons. Thus it must be consulted with a much thicker deletion layer.*

In the "original" planar type Beta-Voltaic nuclear battery shown above, the Beta particles emitted from the radioisotope enter into the p-n junction creating electron-hole pairs. Electrons in the n-region drift across the junction due to the electric field, allowing extraction of an electrical current, much as is done in a solar panel.

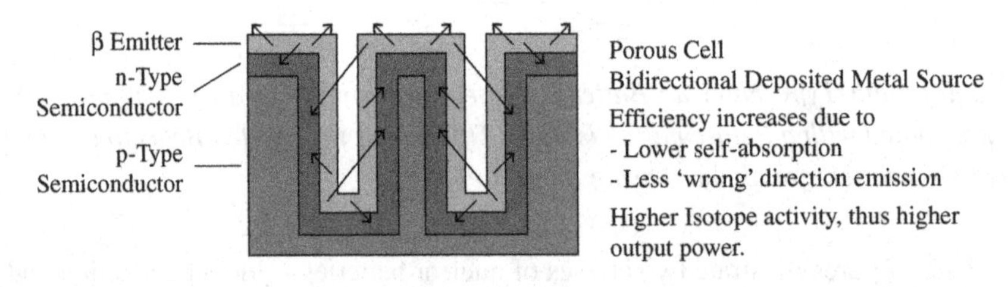


***Porous P-N Junction Beta-Voltaic Nuclear Battery.*** *This design offers a larger surface area per unit volume than in the plasma case shown in the second figure. With the emitting isotope coated on the surface of the pores along with the p-n junction the power density is much improved compared to the second figure. However manufacture of this configuration poses challenges.*

The next configuration shown above greatly increases the p-n junction surface area compared to the prior planar cell. Hence the available electric current per unit volume is

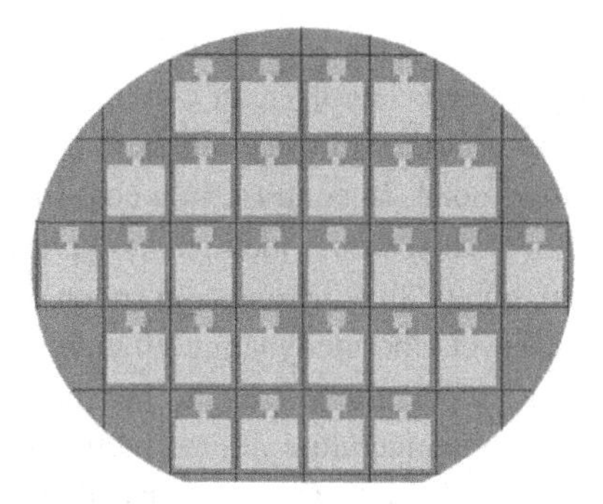

***Microwatt P-N Junction Nuclear Batteries.*** *Manufacturing the p-n junctions for these batteries requires semiconductor techniques. As shown above, many separate p-n junctions are manufactured on a single wafer to speed up the process.*

more competitive. The charged-particle (Beta particles from $Ni^{63}$ decay) collection efficiency is also significantly increased because particles escaping from one side of the isotope layer enter the junction region on the opposite side.

Some of the technology involved in making such batteries is illustrated by the final figure. It shows a number of p-n junction diode units, including gold plated electrical contacts, manufactured on a large round silicon wafer. The wafer is laser-cut into multiple cells $12 \times 17$ mm giving an area of 1 $cm^2$ each. The radioactive source ($Ni^{63}$) is then electrodeposited onto the individual cells to form a nuclear battery. This work was part of Ben Ulmen's MS thesis research (as described in Chapter 14, Ben is now working on the HIIPER plasma thrusters). Later Nie Luo, Research Assistant Professor, made important contributions to the extension of this concept to a multilayer design for use with higher energy Betas from isotopes like Sr-90 extracted from nuclear wastes.

My interest in nuclear batteries actually grew out of a series of events starting in the early days when new nuclear power plants were being installed across the United States. Concerns about the low conversion efficiency of early nuclear power plants leaped into public attention (and also mine) when a sport fishing magazine featured a cover article in the 1970s on industrial waste heat dumped into the Allegheny River near Pittsburgh, PA. The magazine article said the elevated water temperature was killing most fish originally found in rivers around Pittsburgh and other industrial cities. The growing numbers of power plants and mills were located on the Allegheny river to obtain cooling water. Nuclear fission power plants were cited as major culprits. (While not a factor in this article, it is interesting to note that the first commercial nuclear fission power plant demonstration had been the Shippingport Reactor, located near Pittsburgh. It was designed under direction of Admiral Rickover's nuclear submarine office in the Navy.)

The article pointed out that nuclear plants had a lower efficiency than coal-fired plants, hence they ejected larger amounts of heat per unit power level into the river. Why was this? In fission reactors the fuel elements are heated by the fission fragments created by uranium fission reactions in the fuel elements. The fragments lose their kinetic energy quickly in the dense uranium, hence heating the fuel rod. This heat is transferred from the fuel element to surrounding coolant. However, in conventional water moderated/cooled reactors, the central temperature in the solid fuel element is limited to a maximum temperature set by the fuel melting point. In contrast, in a coal-fired boiler, the flame heat is transferred through a thin boiler tube wall which has good heat transfer properties such that the temperature limits are higher. In a thermal conversion cycle like these, the limiting efficiency (the famous Carnot cycle thermodynamic efficiency limit for a heat "engine") is set by the ratio of the cold temperature versus the hot temperature in the cycle. Thus nuclear plants with the maximum temperature limited by the fuel rod melting point had a reduced efficiency versus coal-fired plants. Over time fission reactor fuel element designs have improved, allowing them to go to higher temperatures, hence more competitive efficiencies. Still, cooling any power plant is difficult and requires use of large cooling towers, a lake, an ocean, or a river for the "heat sink." Many sites are excluded for locating plants (no free land, water temperature restrictions, etc.). Thus, in some cases, offshore barge-based systems have even been suggested.

When I began thinking about this problem, I reasoned that we must find a way to directly convert the fission fragment energy into electricity. Such a Direct Energy Conversion (DEC) process would avoid the Carnot cycle efficiency, because DEC is not a thermal process. Thus the use of DEC would maximize the plant efficiency versus using standard thermal energy conversion. After all, the fission fragments are moving charged particles providing an elemental form of electrical current. The problem, however, is that the highly charged fission fragments interact so strongly in solids that they quickly slow down, transforming their kinetic energy into thermal energy associated with the temperature rise of the solid. Then in conventional fission reactors the heat in the solid fuel elements is extracted by having a heat transfer fluid such as water flow by the fuel elements and onto an energy converter like a steam turbine. This approach misses out on the opportunity of directly extracting any of the original electric current.

I decided that one way to achieve direct conversion using the electrical current formed by the fission fragments was to use very thin (micron thickness) fuel layers, such that the fragments could escape the fuel element. The use of a high voltage "collector" surrounding the fuel element could then electrostatically convert much of the fragment kinetic energy into a high voltage electrical output. Like some of my earlier concepts in other fields (e.g., the nuclear-pumped laser), I originally thought that I was first to propose this approach. Thus, I went happily about conceptual designs for such plants, which I termed "fission electric cells." In those days, without use of the Internet, it was harder to do an extensive literature search. However, in a few weeks I found that George Safonov at the Rand Corporation had already proposed this concept. Researchers at NASA's Jet Propulsion Lab (JPL) and the Battelle Memorial Research Institute in Ohio had then picked up on Safonov's work and were expanding the concept, including doing experiments on fission fragment escape from thin films of various thicknesses. While I was initially taken back with this revelation, I rejoiced in realizing there were fundamental physics and engineering issues involved in the concept that I could hopefully contribute to. This Black Swan had eluded me again, but there were still many Mute Swans left to find!

This research eventually led me to consider direct conversion energy from charged particles escaping radioisotopes, and also related methods for conversion of gamma and x-ray emission. Even neutrons can cause proton recoil in plastic materials, providing charged particles for direct conversion. Thus direct conversion of nuclear radiation energy is a "rich area" for varied research. After publishing some research papers about radioisotope powered batteries, I was asked to become a consultant to the Nuclear Regulatory Commission (NRC) on safety evaluations for early radioisotope powered pacemakers. At that time "nuclear" pacemakers were widely used to take advantage of their very long lifetime. Chemical batteries were cheaper but resulted in rather frequent operations to replace them. This trend continued for a few years, but it may be recalled that many of these chemical battery versions were upgraded to provide improved lifetimes and a lower cost. As a result of that, including concerns about keeping track of the radioisotopes to prevent disposal in waste incineration, chemical batteries replaced nuclear ones.

My 1973 book for ERDA on direct conversion of radiation energy (described earlier in Chapter 9) captured much of this nuclear battery technology along with many other concepts. It provided the underlying basics for each as best I could develop it. At that time I was personally thinking about application to power sources. As discussed earlier, I was pleased to find that the many persons working on radiation detectors used this book to help develop improved radiation detectors. Also NASA researchers designing plutonium powered electrical power for space probe applications used it as a reference book. While my work on nuclear batteries continued a few years after publication of this 1973 book, I gradually moved to other areas.

Thus, the book and nuclear batteries were far from my thoughts when results of DOE's first NERI (Nuclear Energy Research Initiative) project awards were announced about ten years ago. The NERI research support was created by DOE to reinvigorate the fission reactor field. I was greatly surprised and pleased to learn that Sandia National Laboratory and its collaborators won a contract to study the fission electric cell described in my 1973 book. Their argument was that advances in materials technology and other related areas could potentially make this reactor possible. If so, it would be a Black Swan development in fission power (my terminology). Unfortunately, this study ultimately concluded that we are not yet up to doing this, largely due to materials limitations. However, with the Sandia work much more is now known about the concept, and my prediction is that its day will come.

Then a second unexpected event happened that brought me back into research on nuclear batteries. I received a phone call one day in 2004 from Amy Duwel, a scientist at the MIT Lincoln Laboratory in Cambridge, MA. She asked if I was the George Miley who wrote the book on direct energy conversion that covered nuclear batteries. After I confirmed who I was she said "Good! We didn't know if you were still alive!" Later as we got to know each other better, we joked about that conversation. She asked if I would consult on a nuclear battery project she headed at Lincoln Labs. It was funded by DARPA to develop nuclear batteries for use by soldiers instead of chemical batteries. Long-lived nuclear batteries could lighten the load carried by soldiers into the battlefield. Hers was one of five teams funded by DARPA to do this study. Amy had proposed a battery using a concept she termed "Post-Activated Secondary Electron Radio-Isotope MEMS Battery." This battery would use a nuclear reactor irradiation after the unit was already assembled to create a radioisotope, e.g., promethium, for its power source. I became quite active on the project and enjoyed returning to the field (should I say because I am still alive?). That project ended after two years but the experience renewed my interest in nuclear batteries.

As the DARPA project ended another development came to my attention. Researchers at the Universities of Wisconsin, Cornell, and Rochester began publishing papers about the application of micro- and nano-technology to small nuclear batteries (radioisotope powered direct conversion units, conceptually like miniaturized pacemaker nuclear batteries). These batteries have very low micro- or milli-watt powers, but are of great interest for use in instrumentation and sensors that require extremely long lifetime and must survive harsh conditions. Applications range from airport sensors in remote air-fields to NASA space probes. The use of nano manufacturing provides a whole new dimension for the design and performance of these remarkable batteries. I became excited reading about these developments and decided to try to get in on the action. After much review, I proposed a unique $Ni^{63}$ (pure beta emitter) powered porous silicon p-n junction type of battery to DOE. This design, shown in the third figure at the beginning of this chapter, uses a porous emitter electrode structure to increase the source surface to volume ratio, resulting in a relatively high efficiency and power density. This research was subsequently funded, and we managed to demonstrate a prototype.

The problem I faced in starting this work was the need to come up to speed in the nano-manufacturing of p-n junctions in this three-dimensional configuration. I had the persistence and optimism to dive into this full force. However, I had to rely on students and colleagues with appropriate experience in the related areas to help. Rich Masel, Professor of Chemical Engineering, had considerable experimental equipment for construction of various p-n junction materials and also nano-structures. Rich used these structures in micro-energy converters for reforming of natural gas to obtain hydrogen. While talking to Rich about this I learned that fortunately he knew of p-n junction type nuclear batteries and was also interested in them. Thus we agreed to collaborate on the project. Rich had P. D. Desai, a graduate student, and Sageed Moghaddam, a post-doc familiar with p-n junctions, assist us. Ben Ulmen, a graduate student in NPRE who had previous experience with semiconductor materials and their production, joined the work on the p-n junction cells for his MS thesis. (Ben is now doing his PhD on HIIPER as discussed in Chapter 14. It turns out that he had an earlier interest in fusors and enthusiastically undertook the HIIPER project.)

Later, Nie Luo, who worked with me on fuel cells, joined to help extend this effort to the design of cells that could use radioactive sources taken from stored nuclear wastes at Savannah River National Lab (SRNL). Government officials are anxious to find uses for these wastes. In addition this would also provide a relatively cheap radioactive source for use in nuclear batteries with large reserve supplies available. The problem, however, is that the emissions from the main waste such as isotopes like Sr-90 are not well suited (the beta energies are too high and some gammas are emitted simultaneously) for nuclear p-n junction operation without major changes in converter plate design. This I opted to consider multiple layer p-n junction concepts such that energetic gammas would interact over the multiple layers. Recently we demonstrated the operation of a prototype multi-layer p-n junction nuclear battery. Preliminary studies were encouraging, but funding ended before we could fully evaluate the concept. Personnel have moved on, so it appears that the concept must wait for interest by other researchers and renewed interest by the DOE to continue.

## Further Reading

G. H. Miley, A. Duwel, "Post-Activated Secondary Electron Radio-Isotope MEMS Battery," *Proceedings, DARPA RIMS Kick-Off Meeting*, Arlington, VA, Oct 14–15 (2004).

G. H. Miley, J. R. Lee, B. Ulmen, "Honeycomb Betavoltaic Battery for Space Applications," *AIP Proceedings*, Vol. 969, pp. 557–569 (2009).

G. H. Miley, B. Ulmen, P. D. Desai, S. Moghaddam, and R. I. Masel, "Long Lived, Low Power Betavoltaic Nuclear Battery for Electronics," *7th International Energy Conversion Engineering Conference*, Denver, CO, p. 4601 (2009).

# Chapter 16

# Computation and Theory

COMSOL Application Mode Coupling For Fuel Cell Analysis

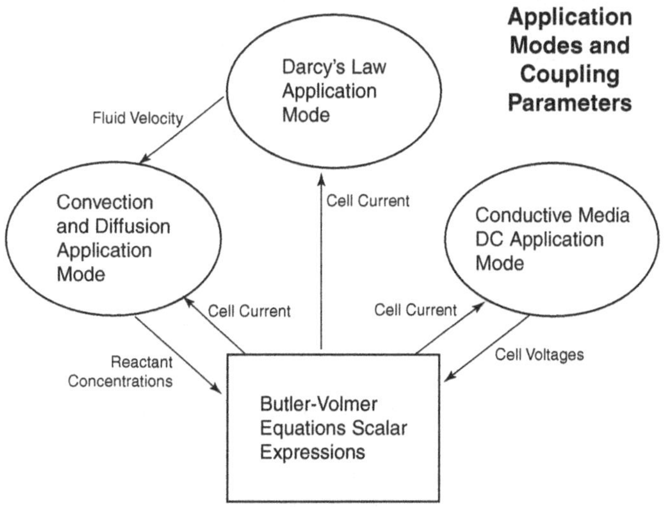

**Application Modes and Coupling Parameters**

Darcy's Law Application Mode
Fluid Velocity
Cell Current
Convection and Diffusion Application Mode
Conductive Media DC Application Mode
Cell Current
Cell Current
Reactant Concentrations
Cell Voltages
Butler-Volmer Equations Scalar Expressions

*As shown here conceptually, COMSOL Multiphysics can pull together and provide integrated convergence of a variety of modules containing the basics physics equations needed for that module. Thus in fuel cell analysis we combined the modules (modes) shown here to provide a comprehensive multi-dimensional solution. The schematic illustrates how Ethan Byrd, a graduate student from ECE who did his thesis with me, set up such a problem. Governing laws for flow in porous media (Darcy's Law), convective and diffusive flows, chemical reaction kinetics (Bulter–Volmer equations), and electrical conductivity for both ion and electron flow (conductive media mode) are solved in a self-consistent manner using the couplings shown in the diagram. Ethan's work was used to analyze results from our fuel cell experiments and then help guide next step designs.*

My undergraduate studies at Carnegie Institute of Technology (now Carnegie Mellon University) exposed me to the "Carnegie Plan of Education for Engineers," which builds on learning from solving open-ended problems — originally stressed by the founder of Carnegie Tech, Andrew Carnegie. This plan also espoused the view that one should bring computations, modeling, and theory along simultaneously with experiments and system synthesis. This approach flies in the face of the current trend towards specialization where one focuses on a single aspect of a problem, excluding others. The Carnegie plan incorporated the philosophy of bringing all the tools/skills that one could muster to find a "best" solution for the problem at hand. It was recognized that generally there is not a unique solution to real world engineering problems. Instead, these problems are open-ended and usually involve tradeoffs to find an "optimum" solution. My determination to follow that route in my work was reinforced quickly during my first job at KAPL. There, as described earlier, I effectively used a combination of the powerful (at that time) main-frame computer resources plus the extensive reactor core experimental facilities available in the Nuclear Navy program. This allowed rapid development of the new concept of "burnable poisons" for core life extension. I continued this approach in my personal research and have always encouraged my students to do so as well. In advising students, I also always try to help them build on their strength (be it experiments, computer simulations, etc.), while also incorporating pieces of complimentary aspects. Thus, I have supervised a number of theses that may appear to be computational or experimental. However, a close examination will reveal that they typically have carried out some rudimentary experiment to compliment a theoretical study, or, in the case of experiment studies, some theoretical/computational work to provide insight into the basic mechanisms involved in the experiment. In my opinion, this approach has strengthened their work and has broadened their perspectives. Several students who have returned to the campus for a reunion after working in industry for some years confirmed that learning this approach put them in good standing in their current work. I have not conducted a poll to be sure this is an "overwhelming" point of view. I just hope it is!

It is my impression that outsiders are a little confused by my approach and my tendency to work in several areas simultaneously. My research has generally favored experiments. However, along with students, I have also done a number of path-finding theoretical/computational studies over the years. I enjoy idealized theoretical development, but practical applications generally require the development of new methods and simulations that can treat realistic conditions, geometry, etc. Such modeling was a cornerstone of my early nuclear fission reactor kinetics studies. I also undertook several involved computational studies of NPLs, dealing either with electron energy distribution calculations for radiation-induced plasmas or estimates of excited level densities and population inversions in various gases during nuclear pumping. Both of my early books on direct energy conversion and fusion energy conversion are rich in original deviations of novel concepts involving charged particle transport on to fusion MHD compression-expansion cycles. The short, but succulent, deviations of the gamma-electric cell for generation of ultra-high voltages were noted earlier. Indeed that work led to intense studies of the gamma electric cell at LANL, which ultimately resulted in a new C-division project started there in the 1970s. I was so pleased when Hiromu Momota, Professor of Plasma Physics at Nagoya University, carefully reviewed the developments in my book on fusion energy conversion and wrote me a long letter with several suggestions for modifications and improvements. Hiromu was, and is, a very highly respected plasma theorist in Japan. I was honored to engage him in discussions about topics in the book. And, as already noted, this resulted in a close collaboration that lasted many years.

Indeed, later after his retirement, Hiromu joined me at Illinois with the goal of being co-editor to a revised version of the book. Unfortunately (from the point of view of the book) we became so involved with FRC and IEC research that the book never got done! But his time at Illinois (almost four years with us before returning, somewhat homesick, to Japan) was very productive. Numerous FRC and IEC papers came out, all with important insights from Hiromu's great mathematical skills and knowledge of plasmas. I also took full advantage of the opportunity to get Hiromu involved as a co-advisor for several forefront theoretical theses during this period. These included Hung Joo Kim's stability analysis of IECs and Linchun Wu's extensive theoretical development of cross sections for heavy ion beam transport in heavy ion beam fusion. They, and other students during this time, greatly profited from the guidance Hiromu generously offered. Hiromu's expertise also aided an important experimental project sponsored by NASA which used an electron source to simulate a proton collimator for use on a $p$–$^{11}B$ fusion propulsion. The logic for the collimator experiment was that the electron version could be reasonably small, but the results could be analyzed mathematically to understand the much larger MeV proton version needed for the real fusion thruster. Hiromu's help was also instrumental to the success of a DOE sponsored study of a "dipole-assisted" IEC. Hiromu has clearly joined the "happy warrior" group in search of the Black Swan. He has had many sightings of Mute Swans and is continuing the search while retired. I take my hat off to him as he continues on his journey.

Most of my research with students and my own theoretical work and computational studies involve plasmas, particularly related to fusion systems. In recent years, however, I became quite interested and involved in use of the commercial code, COMSOL Multiphysics for fuel cell analysis. Later I began plasma studies with it also. This has the advantage that a number of physics modules are built in for use and can be specified for various regions in a 3-D CAD drawing of the system of interest. The code then has powerful convergence logarithms and various mesh patterns that allow amazingly rapid solution of complex problems in realistic geometries. Due to my work using this code, I was asked by COMSOL officials to host a workshop on it at Illinois in 2008 and again in 2012. All chairs were filled in the lecture hall for these workshops. Various company staff and other users discussed recent advances made with this popular code. One area we have made considerable use of it is for analysis of our fuel cell. This involves combining descriptions of fluid flow, heat transfer, ion transport, electrical circuits, and chemical reaction kinetics in a complex 3-D geometry. Because our fuel cell, the direct borohydride-partial cell, is an all-liquid cell (versus the conventional $H_2O_2$ cell), avoidance of two-phase treatments seemed to simplify matters. However, many added complications popped up. Most were linked to the fact that this cell design was pioneering the field, so we had to develop a whole new database of fundamental constants and rate constants. This, in turn, required a series of base experiments designed for specific data measurement (e.g., "half cell" studies to determine the interchange current parameter). Finally, in view of the various approximations involved, careful normalization to an experiment was essential. Once done, the modeling could be used with some confidence in a region that does not stray too far from the normalization point. We presented our results at several national COMSOL meetings where I was pleased to find considerable interest from other fuel cell modelers. In addition, I took every opportunity possible in these meetings to expand my knowledge of COMSOL via discussions with others and participating in short courses on specific topics. In addition to research, I envisioned uses of multiphysics methods in classes and a text on such applications (some COMSOL teaching texts are available, but usually focus on a specific topic area or discipline). Unfortunately, that project has never gotten off the ground due to my time constraints, so I am not sure that it remains in the cards.

More recently we have turned to COMSOL analysis for our HIIPER plasma thruster experiments (for an example, see the graph at the end of Chapter 14). This has been enabled by the addition of a charged particle tracking module and a plasma module in recent releases of the COMSOL software. These modules, combined with the existing RF and EM field modules provided us with the ability to study the complicated HIIPER plasma. Some approximations are still needed, mainly due to the non-equilibrium nature of the IEC plasma. However, COMSOL results from our initial study have already provided insight that will aid interpretation of data from various diagnostics (e.g., a Faraday cup, an Electrostatic Mass Analyzer, and a force plate) employed in the experiment.

## Further Reading

G. H. Miley, D. Driemeyer, and W. C. Condit, "A Monte Carlo Method for Calculating Fusion Product Behavior in Field-Reversed Mirrors," *Computational Methods in Nuclear Eng.*, Vol. 2, pp. 7–37 (1979).

G. H. Miley, J. Demora, B. Jurczyk, and M. Nieto, "Computational Studies of Collisional Processes in Inertial Electrostatic Glow Discharge Fusion Devices," *18th IEEE/NPSS Symposium on Fusion Engineering,* pp. 23–26 (1999).

G. H. Miley, L. Chacon, D. Barnes, and D. Knoll, "An Implicit Energy-Conservative 2D Fokker-Planck Algorithm," *Journal of Computational Physics*, Vol. 157, pp. 654–682 (2000).

G. H. Miley, G. G. Hawkins, and J.A. Englander, "Use of COMSOL to Model the Flow Field of an All Liquid PEM Fuel Cell MEA," *COMSOL Users Conference*, Boston (2006).

G. H. Miley, L. Wu, and H. Momota, "Computational Study of Multi-Electron Ionization in Low-Charged Heavy Ion-Atom Collisions," *ANS Transactions*, Vol. 99, pp. 245–246 (2008).

# Chapter 17

# Nuclear Power Plant Safety and the Illinois Low-Level Waste Site

*View of the Clinton Nuclear Power Station near Clinton, Illinois, about 40 miles from the University. Brought online in 1987, this plant uses a GE boiling water type fission reactor design. Its construction was typical for that period in history — way over budget in construction time and cost. In the early years of operation, Clinton operations received numerous citations by the Nuclear Regulatory Commission for safety violations. Thus new management was brought in to solve these problems. Subsequently I became involved for some years as an "independent" member of their safety audit committee, called the Nuclear Regulatory Audit Group (NRAG). That was a memorable experience for me, and at the same time I feel that I contributed to the effort to correct the situation. Now, over a dozen years after I retired from NRAG, Clinton has become one of the plants in the United States with a highly rated operations record and is considered a low cost electrical power producer.*

In 1988, Don Hall, then manager of the Nuclear Power Station at Clinton, Illinois, gave a seminar at the University of Illinois, which I attended. Don was admired in the nuclear power field, being one of the "hard-nosed" managers now spread across the country who came from the Rickover Nuclear Navy office. Don had been brought in to "straighten out" things at Clinton. During a discussion afterwards, he proposed that we meet for lunch to discuss my possible involvement in the Clinton power station. He said that they had a safety review committee which reported to the Nuclear Regulatory Commission (NRC). This committee was called the Nuclear Regulatory Audit Group (NRAG). Each power station in the United States has such a committee which reviews all operations and provides independent reports to the NRC. The one at Clinton had members from five of the plant's operational departments along with several experienced consultants from Washington, DC. Don wanted to add an independent member who might bring in fresh points of view. He convinced me that I could do the job. I agreed, and this led to fifteen years of service on the NRAG. At the first meeting, I discovered that some of the other members on the committee were not as enthusiastic about my participation as Don was. The fact that I was coming from the University and involved in "exotic" research like fusion gave them the perception that I was not practical enough to really contribute to everyday power station issues. When I realized this, I became determined to prove them wrong. I did that by working hard, reading numerous incident reports from the plant carefully, and frequently talking to people at the plant to formulate my points of view. Eventually it was recognized that I could come up with issues that others there who were immersed in daily operations overlooked. In a way, I was looking for possible Black Swans — in this case, Black Swans that we wanted to avoid, ensuring safe operation of the plant.

Each NRAG member chaired several subcommittees which focused on certain aspects of plant and operation safety. I chaired subcommittees on reactor operator training, water chemistry, and emergency preparedness exercises. I thoroughly enjoyed this work because I felt I was making a contribution which I was uniquely qualified to do, even though the practical issues involved were quite different from those in the academic world. Still, there is a common denominator that my research experiences had taught me; namely, to keep an inquisitive and open mind.

One aspect of human nature became obvious to me early in my NRAG review work. After an "accident" or a "mistake" happens, those involved tend to side-step the real root cause, particularly if it reflects badly on their work. Instead, they bring up all types of side issues. It was my job to cut through this "smoke screen" and uncover the true root cause so that the "lessons learned" would prevent reoccurrences. This required some skill in understanding people and reassuring them that it was to their benefit to assist in this root cause analysis. I felt I was becoming quite good at this, but I was brought back down to earth by a series of events right after a major Black Swan event for nuclear power in the United States — the Three Mile Island accident in Pennsylvania. Because Illinois has a number of nuclear power plants, the governor wanted to reassure people that they were safe and would not suffer a similar accident. He quickly appointed a special task force to review operation of the nuclear power plants. I was assigned the task of evaluating reactor operator skills and training at several power plants in the Chicago area. One operator I interviewed made a good impression on me and I gave him top ranking. However, the week after my visit I learned that this operator had been involved in an "incident." He was following some written instructions and when he flipped the pages in the instruction booklet, two pages stuck together. However, it so happened that the number sequence continued properly such that it he failed to realize that a page was skipped. The incorrect procedure had him holding three buttons down simultaneously. Because they were widely separated, being ingenious, he taped one down and manually pushed the remaining two. This mistake "scrammed" the reactor. That was not a safety problem, but caused the company financial loss because the plant was offline for hours due to the lengthy start up procedure necessary after a "scram." The operator was required to undergo further training. I realized that I had made a mistake in my evaluation. How could a "star" operator not realize something was wrong when taping a control button down was required? Anyway, as they say, dealing with science, despite its unknowns, can be more predictable than dealing with humans! Thus plant operating procedures must have backup actions built in to protect against the possibility of human error.

My duties on the Clinton NRAG subcommittees required me to spend several days a month at the plant reviewing activities in water chemistry, operator training and emergency planning. Water chemistry is extremely important for light water nuclear plants in order to maintain the quality and pH of the water to prevent corrosion of vital components. Part of the problem occurs due to radiation dissociation of the water into hydrogen and oxygen as it passes through the core region. Ideally these products would be recombined to maintain neutrality. However to aid that process, chemicals are added. Two issues develop — short term and long term corrosion. The latter is one of the factors that may limit plant lifetime. This becomes particularly crucial with the lifetime "extensions" requested from the NRC by many nuclear plants in the United States. Many plants are currently operating well past the traditional 20 year lifetime originally assigned to them. Because no new plants have been built for years, extensions have been vital to maintaining the U.S. nuclear "fleet."

*Lifetime* is complicated to define. It involves a combination of technical and economic issues. Think of your car. Many people trade in their cars to get a new style, improved gas mileage, avoid increased repair costs, and added safety features well before the car stops working mechanically. In view of resale value and changes in new car cost, some studies have shown that it might be most economic to trade in a car every 3 or 4 years. Such considerations also enter power plant lifetime determinations. With constant changes in safety requirements, expensive modifications are often needed over the years. On the other hand, water chemistry and corrosion issues can affect extensive lengths of piping as well as components, posing physical issues that may be extremely expensive to repair, setting an "absolute" lifetime. Thus the water chemistry department not only monitors the water quality and pH, but also metal coupons (samples) placed in the water flow. These coupons are periodically removed to check for corrosion. Because the corrosion process is relatively slow, the interpretation over short test periods can be difficult.

My interaction with the water chemistry people during our visits to Clinton mainly focused on reviewing records from measurements of water quality and examination of coupons taken from various locations in the water loop. Because any changes occurred fairly slowly, the challenge was to identify trends that might affect near-term operation as well as trends that might affect long-term plant lifetime. This work did not involve exciting operational issues, so I felt that one responsibility I had was to bring any issues to the early attention of the management and NRAG committee. I think I gained a reputation for always discussing water chemistry in some detail at NRAG meetings. While this subject seemed boring, I convinced people that corrosion trends should not be allowed to creep up on the plant. If allowed to continue too long, corrections could not easily reverse the harm that may have already occurred.

The other two subcommittees, operator training and emergency preparedness, involved much more human activity and easily caught everyone's attention. Operator training required a combination of classroom studies and practice operation on a simulator. New operators went through this training before they could qualify for a license, but in addition licensed operators were required to return for refresher training monthly. The simulator was set up with controls located much as they were in the actual control room, but in the simulator the various plant responses were mimicked by a computer program. I would sit in on classes and observe training on the simulator when I visited Clinton. The simulator training is like the training given to airline pilots on flight simulators. Operation under normal conditions is performed along with operation under extreme emergency conditions. The operator could do things on a simulator that would cause great harm to the actual plant, but of course didn't damage the simulator. The issue, though, as I saw it, was to be sure that the operators felt that accident conditions based on the simulator were realistic. Knowing that this is simply a computer simulation can lead one to be fairly cavalier about what they did in response. One way to help guard against this is to be sure that the events leading into the off-normal or accident conditions come about in a manner so that the operator does not anticipate what is going to happen in advance. Thus off and on I also spent time with the personnel programming the simulator computer for the next day's training exercises. Fortunately, Clinton had excellent personnel in their pool of operators, and I usually ended up with very positive reports about the quality of their training.

All nuclear power plants are required to undergo a drill for emergency preparedness every few months. These drills were exceedingly complex, involving not only the plant personnel but also local emergency preparedness units such as police and medical staff. The drill was performed simultaneously while the plant was still operating. Thus the people participating in the drill were housed in a separate building in a room outfitted with extensive communications capabilities. About half of the plant staff were involved in the drill, while the other half maintained plant operation. Thus all announcements made about the emergency conditions had to be preceded by the message "This is a drill." The formulation of the events that occurred in the drill was composed by a group of about five senior personnel at the plant. Planning the drill is something like composing a play. Usually the drill involved initial conditions that gradually developed into a major problem that the personnel participating in had to stabilize.

Emergency preparedness teams outside of the plant were brought in through various scenarios. For example, a fire might be reported in a certain location and fire trucks from the neighboring town of Clinton were called to respond to help the plant firefighters. Going into a fire in a nuclear plant raises unique conditions that are not encountered in normal fires, such as possible radioactive contamination. The fire responders had been trained to handle such conditions such that they would avoid becoming contaminated, or if they did, to rapidly hose themselves down to get rid of it. Another frequent part of the exercise would be for one or more workers to be injured, requiring their removal to the local hospital in Clinton. Often the drill scenario would have one of the "injured" workers become badly "contaminated." The emergency room personnel then had to do a decontamination before working on the patient. The staff had been trained for such events, but these practices reinforced that training. If the drill assumed radiation leakage from the plant, crews were sent out to pick up samples from collection stations located at various distances from the perimeter of the plant. This data, along with meteorological data, provided information about any radioactive plumes that might be carried into neighboring areas. Neighboring communities all had warning sirens mounted in appropriate places and these would sound off warning of a possible radioactive plume approaching their area.

NRC inspectors were present at these drills, and filed lengthy reports on the performance of the plant staff and other responders. The first several drills I attended received roughly a B-grade, but continual training in between drills eventually got that up to an A. My function was to observe the drill and provide feedback to management and the NRAG committee on suggested ways to improve the drill responses. As an observer I wore a white armband to signify that I was not participating in the drill and that participants should not discuss anything with me. Also, I was not to speak to them. In the middle of an intense situation created by the drill, it was sometimes hard to avoid saying something. I must say that for awhile I felt a little out of place standing in the midst of a simulated disaster such as a fire and not doing anything while others were running around me trying to stop it and save "injured" people.

My NRAG activities also put me in close contact with people in the Division of Nuclear Safety in the State of Illinois Emergency Management Agency. This agency has certain regulatory authority delegated through the Federal Nuclear Regulatory Commission. Their coverage includes certain aspects of nuclear power plant operations, nuclear waste disposal, nuclear and x-ray sources (e.g., dental and medical x-ray machines, and also industrial radiation sources), and other radiological issues like radon in homes. In the late 1990s the Division of Nuclear Safety in the State Emergency Management Agency was charged with finding a low-level nuclear waste disposal site in Illinois. Subsequently they asked me to chair an advisory committee on technical issues for site selection. I agreed to do this, although later I discovered that most of the waste site decision makers were focused on non-technical aspects and political issues rather than what this technical committee did.

The first very challenging task for the Department of Nuclear Safety staff was to find a location for the disposal site. The State Geological Survey identified some potential sites that appeared to have suitable geology for the disposal facility. Next the local citizens in those areas were asked if they were interested in hosting the waste facility. After much discussion, only one area near Martinsville in southern Illinois (not on the original list of possible sites) agreed to be evaluated for its location. (Later when final voting about the site came after much more study of it, opinions were split. Martinsville residents again voted to accept the waste facility, but other residents of the surrounding Clark County did not want it! The Governor of Illinois issued an order that the final decision on whether or not to proceed with the use of a site would be made by a hearing presided over by a special judge who was appointed by the Governor.) Martinsville was not the best location from a geological standpoint, but the other better locations had not expressed an interest in hosting the site. Thus the Department of Nuclear Safety, headed by Terry Lash, initialed a detailed study of the Martinsville site. Many of the technology issues involved were concerned with the site's geology, which was put under study by the State Geological Survey and other specialized contractors. The basic issue was whether or not leakage from stored waste containers did occur in some way, and could it get into the ground strata and ultimately diffuse or transport into drinking water sources. To study this, the geologists took numerous samples from boreholes to obtain data about the subsurface rock formations. Also, dye injection studies measured water transport through porous regions of the subsurface. All of this data was to be inserted into a very large three-dimensional water transport computation or "simulation" designed to evaluate possible transport of any leaking isotopes from the storage area to water wells used by the surrounding towns.

These computations were started but then delayed because the consulting firm involved kept coming up with requests for more data that required core drilling in the field. Further, the consulting firm said that with the "limited" data at hand, preliminary computational runs would not make sense. This computation was a key input needed for site recommendations by the technical advisory committee. As its chairman I was becoming increasingly impatient. The situation grew heated as frequent requests by newspaper reporters started coming to me demanding predictions about what the transport study might conclude. Thus I decided to take matters into my own hands and undertake an independent study. I did this with the help of Kelvin Kuelske, a graduate student who performed the work as part of his MS thesis. I am not a geologist, but I felt confident that the basic fluid flow analysis, which I do know, could be adapted to this problem. Water flow in an aquifer is a special case of flow through porous media. Treatment of the radioisotopes and their decay processes during the flow are nuclear engineering issues that I am quite familiar with and could add to the transport treatment.

The first task then was to find a simplified computational tool to do this study. I chose PSpice, which is a well-known electrical engineering circuit analysis program. The flow problem could be translated into a series of resistors, filters, and capacitors. For example, the resistance for flow through porous media could be translated into electrical resistance in the model. This would essentially be an O-D model, but it still should provide initial insight into the situation. Indeed, the analogy worked out better than I expected. Kelvin and I had some data from the available core drilling that had been reported by the State Geological Survey to determine input porosity values. We ended up getting preliminary results for this problem months ahead of the full studies by the consulting firm. This allowed me to discuss the problem with reporters on an informed basis, stressing that the PSpice results were preliminary "first" estimates. (In fact, later when the full 3-D computations were released, the PSpice results were shown to be reasonably good, capturing the general trends well.) In addition to this immediate use of the PSpice results to provide some first insights into the waste site situation, Kelvin Kuelske and I presented a paper on the new method at a professional meeting. That presentation and paper about water movement drew interest by others seeking simplified analyses of such problems. This experience illustrates the point that if we don't back away a new problem merely because we haven't had prior experience that directly applies, we may be surprised at what new solutions — and what new approaches — we can develop.

As a matter of interest, I would add that the isotope transport studies showed that the underground aquifer situation at the proposed site allowed a slow transport towards water wells. The time delay for isotopes to reach the nearest well was significant but not as long as might be desired ideally. The Illinois Division of Nuclear Safety argued that the multi-defenses against leakage (isotope packaging, surrounding containment building, and slow transport) made the site suitable. However critics were vocal. The split vote between the city and the rest of the county, although numerically positive due to the larger population in the city, was not viewed by the state legislature as sufficient to approve the location for the facility. The court hearing set up by the Governor was then held on the suitability of the site and the presiding judge too rejected it. Terry Lash, Director of the Division of Nuclear Safety, later resigned in frustration. No other sites were investigated and Illinois still does not have a low-level nuclear waste site. Earlier, the U.S. Congress had passed a federal law requiring that all states form consortia with one or more other states and develop such sites. Illinois and Kentucky formed a two-state consortium, and Illinois had agreed to be the waste site. Thus Illinois and Kentucky were (and remain) in violation of federal law, as are many states across the country. However, other than occasional notes to the governors of the states warning that they are in violation of federal law, nothing has been done about this. Thus low-level waste from hospitals, industrial users of isotopes and power plants (mainly contaminated clothing in power plants) is shipped out of state to locations such as Barnwell, SC; Richland, WA; or Andrews County, TX, that still accept incoming waste. These sites, with continual urging from citizens in surrounding areas, are gradually taking the position that they do not accept nuclear wastes from outside of their region. Their argument is that they do not want their state to be a "dumping ground" for nuclear wastes, part of the "not in my back yard" or NIMBY syndrome. Ultimately this will pose a real dilemma for the many users of radioactivity and for power plants in Illinois and elsewhere.

After my involvement in the waste site committee ended, the State Division of Nuclear Safety recommended me for appointment to the Governor's Advisory Board for Nuclear Safety. I retain that appointment today, having survived reappointments by a string of governors over the years (appointments to such specialized committees have less turnover after elections than more political appointments). As part of this activity, I was asked to organize several meetings of medical and industrial radiation source and radioisotope users in Chicago. It turns out there are surprisingly large number of such users in the state (involving x-ray facilities, electron beam machines, colbalt-60 gamma sources, and various radioisotopes). These radiation sources and radioisotopes are used in a wide variety of ways, ranging from food processing to treatment of glasses to create colored beads for jewelry. The objective of the meetings was to better understand their views about how current safely regulations and inspections affected their operations and also their handing of radioactive waste. We asked representatives for any suggestions for improvements to regulations that might help them, yet retain the high level of safety needed. Some excellent suggestions were obtained and ultimately implemented after these meetings. In addition, the meetings allowed the users and regulators from the Division of Nuclear Safety to get to know each other better.

A year after I was appointed to the Governor's committee, I was joined by David Miller. David had been at the Clinton Power Plant, and had worked on several NRAG subcommittees with me. David and I worked well together, and I later proposed that he become an adjunct Professor in our Department at the University to supplement the Health Physics offering. This association has been very productive and continues today with David running the North American Technical Center (NATC) and Information System on Occupational Exposure (ISOE) through an office in our building. This center compiles data on radiation levels and worker exposure at power plants in the United States and elsewhere.

Being on the Governor's committee has made me very conscious of radiation exposures. I get my main exposure from dental and medical exams and from frequent air travel, both in the air and at the homeland security inspection stations in airports. Still, those are nothing like the old days (prior to the fall of the Soviet Union) when the x-ray machine in the Moscow airport completely blackened the film in my camera (one way to prevent photos from leaving the country). Even earlier when I was around 8–12 years old, I used to enjoy watching my toes wiggle in the x-ray machines used in shoe stores to help size shoes to the foot! No one ever tried to limit the time these were used unless another customer happened to be in line to use it. Another such problem is that some dentists and doctors instinctively take x-rays in a routine fashion without considering exposure issues. All nuclear engineers and nuclear plant workers have been taught the guiding principle that radiation exposure is to be kept "As Low As Reasonably Achievable" (ALARA). Doctors and dentists are not taught this philosophy. Sometimes I have questioned the necessity for an x-ray, but usually the dentist or doctor has resisted changing their procedures. When I first came to Illinois, I found that Marv Wyman, then the chair of the Nuclear Engineering Program, already had some concerns about x-ray exposure, and decided to wear a radiation film badge he was issued for the TRIGA reactor to his dentist's office. When the dentist saw it, he refused to proceed with the x-ray until the badge was removed, saying his machine was completely safe and he resented this intrusion by a patient. Marv got up and left the office to find another dentist. Of course, times have changed, and now periodic inspections by personnel from the Illinois Division of Radiation Safety ensure exposure safety when they visit to relicense x-ray machines. Plus the machine designs have drastically improved, not only in picture quality but also by greatly reducing the radiation exposure given the patient and the operator.

Being on the state committee has also put some extra pressure on me at the University. The Division of Radiation Safety also oversees safety of radiation sources at the University. My IEC and nuclear battery labs come under their jurisdiction which is delegated to the University's Health Physics office. Thus it is embarrassing for me, as part of the oversight committee, to receive violations for operations in my lab. Still that has happened on occasion! For example, some students took their radiation badges (issued for work in my lab) home by mistake. Prolonged exposure to sunlight and florescent lights caused the film to darken. Thus a report came back that the students had been irradiated at high levels, and it required some time and detective work to straighten that out. Another time a student placed a small container housing a low-level radioisotope in a lead block "cave" (for shielding) and located it under a bench in the corner of one of my labs. Then he graduated and left, forgetting to remove it. The bench hid the cave from sight so it was there for some time before another student found it by accident. All radioactive materials in the lab were to be accounted for at all times and this one was obviously not. Thus the student reported the incident to the Health Physics office. No one was irradiated, but having a source, even a very low-level one, lying around unclaimed is not good! I was embarrassed!

Fortunately for me, the main person overseeing my labs from the Health Physics Office is Paul Safranek, who I knew from my NRAG days at the Clinton nuclear power plant. Paul was in the health physics area at Clinton before coming to the University of Illinois. We know each other well and work well together. Paul has gone out of his way to help me ensure safety in my lab plus managing a variety of isotopes. The strangest experience occurred some years ago in my LENR lab. While on a trip, I received a call from my office that one of the workers in the lab had been irradiated. At the time, we were struggling to get any detectable nuclear reactions from the LENR experiment, much less one intense enough to irradiate someone. I called the lab and asked to speak to the student. As it turned out, he was quite emotional in general, and now insisted that he could "feel" radiation. He was not wearing a personal radiation detector because that was not required in the LENR lab. The lab did have several area detectors (just in case!) but neither had detected anything. Thus I doubted he had received a dosage, but the issue still needed resolution. The concept of "feeling" radiation is certainly not something we teach, although there have been reports of people who stoutly claim that unusual ability. Hearing this I was both relieved (I then felt this was not a true radiation exposure event) and concerned about the student, who seemed irrational. So I insisted that he report the incident immediately to our Health Physics office. They in turn sent him to our University Heath Center (mainly to discuss things with a psychiatrist) and also issued an order that he not be permitted to work in my lab which came under their jurisdiction for handling radioactivity. They thought his "lack of understanding of radiation" could lead to safety problems. I am not sure if the way this unusual situation was handled was completely appropriate, but it did eventually resolve the issue when the student left to work in a different lab. I can only conclude that interactions with our fellow mankind often take one into unusual situations. This was not the only one, nor the last.

A related activity mentioned earlier in Chapter 10 involved my preparation of the nuclear environmental hazards and safety analysis report for the National Ignition Facility (NIF) under contract with Argonne National Lab (ANL). This report thrust me into controversy when persons living in the area protested the construction of NIF due to their concern about possible tritium leakage hazards. They claimed that property values would drop drastically. My report had found that tritium leakage was not a concern due to the small quantities involved plus triple physical safety "barriers" that would prevent escape from the target chamber or target manufacturing lab. Again, as happened during the controversy over where to locate the Illinois low-level waste site, emotions and rhetoric took front stage and it was almost impossible to get the protestors to listen to or try to understand technical facts. The challenge for us in the scientific and engineering fields is to cut through these emotions and get the arguments back on quantitative grounds. That is a real challenge in cases involving nuclear energy or radioactivity because emotions run so high when those topics are involved. In the case of NIF, the NRC finally gave approval to move ahead with construction. Thus some protestors gave up, but a few hard-core persons remain to express their views whenever the chance permits.

## Further Reading

G. H. Miley and K. J. Kuelske, "Preliminary Ground Water Flow Analysis of Potential Nuclear Radioactive Waste Disposal Sites Using Electrical Circuit Analogies," *Second Nuclear Simulation Symp and Mathematical Workshop*, Munich, Germany, Oct. (1990).

G. H. Miley, *et al.*, "Capabilities Needed for the SSM Program and the Technical Role, Design Options, and Planning Process for NIF," *ANL Report*, Vol. 2150, pp. 1–21 (1999).

# Chapter 18

# Teaching, Education, and University Administration

*In 2009, we celebrated the 50th anniversary of the Nuclear, Plasma, and Radiological Engineering (NPRE) Department at the University of Illinois. The Department initially started as the Nuclear Engineering Program, a graduate-level interdisciplinary program offering only graduate degrees. This program was then renamed the Nuclear Engineering Department when an undergraduate curriculum was added. The name was later changed to NPRE to indicate its broad scope. Here, Jim Stubbins, current Department Head, is in the center of the photo while (from left) Dan Hang, Roy Axford, Barclay Jones, and I celebrate the unveiling of a departmental name plate on the patio in front of the Alice Campbell Alumni Building. The four of us, Dan, Roy, Barclay, and I came to Illinois in the early days and all are now in our 70–80s. Dan, the oldest, retired about 10 years ago, but the other three of us remained full-time faculty until I retired in August 2010. Several years ago Roy proclaimed himself, Barclay, and me the "Three Musketeers." I stay very active in the department, teaching and doing research despite being retired. Thus Roy and Barclay still allow me to be a proud member of the Les Trois Mousquetaires! Barclay retired in 2011 so only Roy remains full time.*

My views on teaching and education are rooted in my own early education. My teachers in both grade school and then high school were generally focused on the fundamentals and provided me with all of the tools needed to do well in college. My last two years in high school were at a well known prep school (Mercersburg Academy, in Mercersburg, PA) which generally focuses on sending students on to Ivy League schools. However, I had always had my own sights on Carnegie Institute of Technology (CIT, now Carnegie Mellon University) in Pittsburgh. Thus I was very pleased when I gained early admission when the CIT officers interviewed me at Mercersberg just before my senior year.

In regards to my teaching philosophy (discussed earlier in Chapter 16 in relation to my computational research), I was thoroughly grounded in the then-famous "Carnegie Plan" for science and engineering education. The plan was simple enough — the engineer was charged with first defining the problem (said to be the most difficult and often overlooked task in any endeavor). Then the problem as defined was to be broken down into its elements, and each was tackled with the appropriate tools, whether mathematical or symbolic. Environment and economics were an integral part of the attack on the problem. (Now this is a widely accepted principle in view of the present energy and environmental crises. However, in those days these issues were just beginning to be taken seriously.) We were taught that there was no single right answer. Solutions were often a compromise and involved a trade off and optimization of all factors involved. While well defined, the Carnegie Plan posed a real challenge for instructors who were to give homework and exams that fit into this model (i.e., contain a reasonable number of problems that had "open ended" answers). Some of my instructors clearly struggled with this because single-answer calculation-type problems are easier to make up and certainly easier to grade. However, everyone at Carnegie Tech in those days made an effort to do this to some extent. Ultimately, the Carnegie Plan approach became a part of my perspective on attacking my own research and teaching. These principles are, I think, clearly visible in my career.

My teaching career began with a memorable experience. I encountered a personal challenge in my first class on fission reactor physics at the University of Illinois in 1961. At that time, the military was sending many armed forces members back to universities for added training. Nuclear engineering was a popular field then for all services, and the army was involved in a mobile fission reactor project to fly small power reactors into remote locations. The naval submarine and aircraft nuclear power programs were also continuing to grow. The Air Force had a project aimed at a nuclear powered airplane. The Army and the Air Force projects were short lived when fears about reactors crashing in populated areas caused cancelation of funding for both. However, the nuclear navy remains a major program and a success story. In view of this background, it is not surprising that fifteen of the twenty-five students in my first reactor physics class had been sent to study at the University by the Army. They were generally lieutenants, but one was a Lt. Colonel. I quickly learned that he thought he was in charge of the class — giving orders to the other students and also directing me as to what he thought I should teach. When I realized this, I was equally determined not to take orders from him. Thus, we had a gentle "tussle" during the classes at the beginning of the semester. But fortunately we grew to understand and like each other. I discovered that our bickering about topics needing to be stressed actually had the beneficial effect of getting the other students engaged and thinking broadly about where I was trying to go in my lectures. What I feared might be a teaching disaster turned out to get a high rating from all, including the Lt. Colonel! I learned something about teaching that stuck with me. In the modest size classes that I often had at the graduate level, I typically try to get the class involved in discussions by asking questions. Such an interactive atmosphere is, in my mind, much better than giving a polished one-way lecture. The interactive format is hard to use sometimes. One must be prepared to make the best of a question and not be afraid to veer from the subject at hand so long as the detour provides new insight.

An early example of such a detour happened in the same reactor physics class some years later during lectures on reactor dynamics. A student asked me, "How rapidly can we pulse the TRIGA reactor?" I responded that I did not know, but arranged for the class to meet at the TRIGA reactor several days later where we would do some experiments to find out. I had a reactor operator's license for the TRIGA, and I filed some simple operations paperwork to get approval from the local oversight committee for fast pulsing experiments. Today, with the many regulations, such a request might take months or more for approval because this experiment had not been explored before. The pulses were initiated by rapid removal of a pneumatically operated control rod. The pneumatically operated pulse rod could be "fired" out on a sub-millisecond time period, but the reset and fuel element cool down time between pulses would require minutes. The TRIGA could be run at kW levels steady state and then pulse to a 100-MW peak power with a pulse full width at half max of ~ 20 milliseconds. The large negative temperature coefficient of the special ZrH-U fuel elements automatically turned the pulse around and insured complete safety. However, to achieve full cool down of the fuel elements, a time of 20 minutes or more was traditionally allowed between pulses. Thus I proposed a series of pulses with time intervals between them of order of only a few minutes and challenged the class to predict what would happen in advance. I also made some predictions myself. However I was wrong (as were the class members) in that we did not realize that the negative temperature coefficient of the fuel would cause the second pulse to have a much lower peak power than the first (due to the elevated temperature of the fuel elements following the initial pulse). Another surprise was that the third pulse was higher than the second, and the oscillation of amplitudes continued between pulses, but then died away as the series assumed asymptotic amplitude, which was about 75% of the initial one. In retrospect, it was obvious that the short cool-down time between the first and second pulse caused the lower second pulse amplitude, which in turn deposited less heat in the fuel. Hence, by the start of the third pulse, the fuel was cooler than at the start of the preceding one, so the third pulse had higher amplitude. This pattern continued but eventually damped out as the heating–cooling came to an equilibrium by the time a few dozen pulses were done.

In some respects this could be termed a Black Swan event — something no one anticipated but afterward "experts" quickly claim it *was* anticipated. I thought I knew much about reactor dynamics, but I had not foreseen the pulse oscillation in amplitude. In retrospect, once I observed the pulse variations, I quickly realized how the temperature variations had caused this phenomena. In truth, I was excited, and also very humbled. (Let me assure the reader that the strong negative temperature coefficient in the TRIGA ensured this was not a safety problem even though the detailed behavior of the pulses was unexpected.) Operators are not allowed to do experiments that they do not fully understand. If this happened today, I would have no doubt been in serious trouble with the regulatory people when the operation was reported to NRC regulators who periodically reviewed the TRIGA log book. However, when the review was done, the regulatory official was simply interested in this "crazy" experiment and called me to discuss it.

The students in my class liked the experience of this "dramatic" demonstration and became even more interested in reactor dynamics. Harold Kurstedt, the student who originally asked the question, went on to do his thesis with me on TRIGA reactor dynamics. As part of his work, he employed a second low power TRIGA fuel element core (named LOPRA) in the bulk shielding tank, a second pool of water next to the main reactor pool. A graphite moderator block "window" placed between the two pools allowed neutrons from the TRIGA to diffuse into the outer pool of water where they could irradiate experimental materials or be used to study neutron penetration through shielding materials. LOPRA was then "coupled" to the main TRIGA by neutrons streaming through this graphite column, providing a facility to study the kinetics of coupled reactor cores under a variety of conditions. Earlier I had obtained some "slightly warped" fuel elements from the pulsed TRIGA at the Army base in Bethesda, MD. They were removed from that reactor because their structure could no longer stand stress expansion during high level pulsing, but the elements could run at the low steady state and modest pulse powers involved in the LOPRA experiments. My first PhD student, P. K. Doshi, was instrumental in helping set up the LOPRA and had done some unique neutron pulse propagation experiments using it. This arrangement was now used by Kurstedt to do exciting research on reactor dynamics and also study safety aspects of coupled core kinetics. The LOPRA could run "standalone" at low levels or be driven by neutrons from the main TRIGA. When the TRIGA was pulsed, the LOPRA would perform a much smaller pulse delayed by the millisecond time required for neutron diffusion through the graphite block separating the cores.

*I was presented a plaque showing photos connected with my research areas when I retired in 2010. Mike Reilly, my most recent PhD student who worked on helicon plasmas, is on the left, while P. K. Doshi, my first PhD student, is on the right.*

An interesting historical note is that the LOPRA was licensed by the Nuclear Regulatory Commission, as was the main TRIGA. Thus, Illinois become the first university to have two licensed research reactors. Later, when the TRIGA was shut down and decommissioned, Illinois also became the first to have two reactors decommissioned — a dubious honor.

When P. K. Doshi and I designed and built the LOPRA, we placed the control electronics on a table next to the bulk shielding tank. The control rods were also driven from a control motor connected to steel cables that ran through the tank water to the control rod in the core. The structure holding the core was on wheels such that the whole LOPRA assembly could be moved at various distances from the face of the graphite column to provide various degrees of neutronic coupling. When all of this was ready, a NRC license inspection team visited us to review the setup and corresponding paperwork required for licensing. This was a tense time because I was not sure what the inspectors would think about this unique system. I was pleased when they left agreeing to sign the license provided we would take the controls off of the table and mount them in a panel like "all nuclear reactors used." Getting a license today would be much, much more difficult and I am not even sure it would be possible in view of the present tight regulatory structure. In any case, the LOPRA proved a great research and teaching tool. It is remembered fondly by several generations of students. Its "birth" came as the result of students asking me to explain the Cherenkov glow observed at the face of the graphite column connecting the main TRIGA tank and the bulk shielding tank. Interestingly this distinctive glow had a small but discernible time delay as the Cherenkov glow in the main TRIGA water tank. This led me to consider issues about neutron pulse propagation through the column and eventually to building LOPRA to study these effects in terms of coupled reactor kinetics. Some great things can happen in unexpected ways. This experience is one of many that further enforced my view that creating an interactive atmosphere in teaching is one of the most important elements for effective learning.

Harold Kurstedt went on to become a well known Professor at Virginia Polytechnic Institute (VPI), my father's old school. He is now retired. We enjoyed reminiscing about his LOPRA reactor kinetic experiments when Harold attended the celebration of the 50th anniversary of the founding of the Nuclear Engineering Department at Illinois. We took the opportunity to visit the TRIGA building (the reactor is now closed with the fuel components completely removed). In the photo we are standing on the balcony overlooking the now empty tanks that once contained the TRIGA and LOPRA reactor cores. While seeing this once triumphant site vacant deeply saddened both of us, we managed a slight smile for the photo.

Several recent events caught my attention and made me think that my teaching style of involving students in the learning process (rather than simply presenting material to them in one-way lectures) is also viewed by some as the preferred way to teach public school students. Ruth Bettelheim, a psychotherapist, educational consultant, and writer in Los Angeles, had an interesting article in *USA Today* on November 11, 2010, titled, "With Their Assembly Line Approach, Public Schools Fail Our Kids." She criticizes a passive environment stating that, "Instead of being told facts, children should learn by acting on instructional materials, experimenting and observing until answers are found." Indeed, that has been my goal also. I guess it works at all levels of teaching. But unfortunately, our current generation of youth has become addicted to watching performances on television and other entertainment venues so that they want to be entertained, not involved, in the classroom. It is difficult for teachers to overcome that inertia. A week after reading the Bettelheim article, I attended part of an exciting meeting, "Engineer of the Future 3.0." The motto of this conference series is, "Unleashing student engagement in and for the transformation of engineering education." Daniel Pink, author of the *New York Times* bestseller, *A Whole New Mind and Drive,* gave a talk on "Drive: What the Science of Motivation Can Teach You about High Performance." He concentrated on ways to teach the engineer of the future to have this "drive" to learn. He maintained this requires "gaining the active involvement of the students in their learning." Many influential individuals and faculty from universities across the United States and also other countries attended that meeting. I feel the approach which was encouraged by Pink and others is quite consistent with what I have been trying to do in my teaching over the years. One difference is that I just naturally "fell" into this style, influenced perhaps most by experiencing the Carnegie Plan at CIT. The story about how research on repetitive pulsing came out of class discussions is just one example of this in action. Thus I am excited that more people are recognizing the importance of this approach for education at all levels. However, as I said earlier, the desire of many students to simply be entertained by lectures poses a real challenge.

My office staff at Illinois has also played a tremendous role in my work throughout my career. They not only helped me with the workload (generally well over 40 hours a week), but also provided great encouragement. When I first arrived at Illinois, Carol Mattis was the Nuclear Engineering Program Office Manager under Marv Wyman. Carol knew where to locate everything (department records, forms, etc.) and from long experience she also understood the "ins and outs" of the university system. Plus, she very effectively interfaced with students and staff seeking administrative help. Later, when I became Chairperson of the Nuclear Engineering Program, she continued in this key role. Plus, Carol countered my pack-rat mentality. I vividly recall her annual "office cleaning" where she combed through all files and shelves, tossing many useless documents, letters, etc., that I had accumulated. Chris Stalker took over Carol's role later and continued this great tradition. Chris' husband, Lynn Stalker, was in charge of our three-man machine shop. He could machine fine instruments, do electronics, and repair many things (e.g., control rod and drive equipment for the TRIGA reactor). This husband–wife team made great contributions to the program over the years. Other staff in the business office, student records, admissions, alumni affairs, etc., have been very dedicated and personable, enabling us to "advertise" the Nuclear Engineering Program as having advantages of the large University but being a friendly, small group who knew and enjoyed each other. Indeed, this advantage helped attract the outstanding students and staff we recruited over the years.

After a dozen years I left the Program Chair position to return to being Professor and Director of the Fusion Studies Laboratory. (Note that at Illinois, a program has a "chairperson" while a department has a "head.") At that time I was also editor of three major professional journals (*Fusion Technology* of ANS, and *Laser and Particle Beams* and *Plasma Physics* of Cambridge University Press). Celia Elliot became my office manager. Celia had a most unusual background — besides being an English major, she had been the first female meat salesperson and owner of a meat packing facility in Illinois. Celia and I made a great team. Usually I revised my papers/letters several times, trying to clarify language. Celia said she would do that for me and she did. We often had friendly debates over writing style and grammar, but Celia usually had the last word. A number of authors who submitted papers for the journals I edited were international and Celia helped them correct their English. That was something she was not required to do, but she enjoyed helping them. Also, I ran several major fusion meetings (the Laser Particle Interaction series in Monterey, CA, and the IEEE SOFE which we hosted at Illinois). Celia helped with meeting arrangements, processed registrations, and assisted with preparation of the proceedings. She especially helped with the Russian delegations who attended, picking them up at the airport, making sure they were settled in rooms and that they had their presentations ready in a form compatible with our projection equipment. In those days, just after the collapse of the Soviet Union, the Russians did not have money to attend international meetings. As happened at many professional meetings in the United States, I wrote proposals to various government agencies to get money to support their attendance. This was viewed favorably, because U.S. scientists could learn more about the Russian work. It also served as part of a broader program in the U.S. State Department aimed at discouraging former-Soviet scientists from leaving Russia for positions in unfriendly countries who were trying to develop nuclear capabilities.

Later, Celia took a job as Administrative Assistant to the Head of the Physics Department. Due to many contacts she had developed with the Russians while working with me, she became in great demand to help the Russian scientists write journal papers and proposals. Indeed, Russian administrators ended up paying for a number of trips for Celia to visit their labs (many in the former secret cities) and helped write proposals to obtain research support. These proposals were often sent to U.S. and international agencies set up to fund former Soviet weapons scientists to work on non-weapons topics related to their basic backgrounds. The intent was to keep them in Russia while directing their research away from weapons. Her help was greatly needed since the Russians had never had to write proposals before and did not understand the basic principles of preparing a good proposal. Under the Communist regime scientists had always been assigned research tasks and then automatically provided the money needed to carry them out. So this proposal writing was entirely a new experience for them.

More recently, English majors Autumn West and Robyn Bachar have also worked in the office with me and have continued Celia's tradition of doing great editing of many papers and reports, plus proposals. Robyn has been instrumental in the editing of this book.

I learned much about teaching, research and life in general from colleagues at the University of Illinois. My contemporaries, age-wise, in Nuclear Engineering were Barclay Jones and Roy Axford. Barclay was a graduate student when I arrived as Assistant Professor. However, because of service in the Canadian Air Force prior to returning to university studies, he was actually several years my senior. In addition to many professional and social interactions, I worked closely with Barclay when he served as Associate Chair of the Nuclear Engineering Program during the second half of my tenure as Chair. He was instrumental in creation of the undergraduate program in Nuclear Engineering which enabled conversion from a "Program" status to a Department. Then he served as the first head of this "new" department. One social aspect of University of Illinois life that Barclay, his wife Becky, Liz and I have enjoyed over many years is owning season tickets for seats together on the upper balcony of the football stadium (the upper balcony is exposed to the weather, but provides a great view for seeing plays develop). Off and on over the years I have taken visitors and students, often to introduce international students and visitors to our American tradition of college football. Most are familiar with soccer, but fail to understand American football. Still they enjoy the "show," including the Illinois marching band at half time. As Roger Hancox, my colleague from the UK Culham Fusion Lab said when he was visiting and I took him: "This game makes no sense to me at all. There is no plan — just pushing and tackling each other." European soccer and rugby were his loves, and he thought their play was much more organized and deliberate.

Roy Axford joined the faculty several years after I did, being "lured" away from a key position at Northwestern University by Marv Wyman. Roy had already established a record as an outstanding teacher, starting at Texas A&M and then Northwestern University. At Illinois he has continued this tradition and received teaching awards from our Department, the Engineering College, and the Graduate College. He told me that his technique was to start from basics, have a well-defined goal for each lecture, and deliver the material from "scratch" without notes. His lectures often include notoriously long equations, derived in "real time" and filling all available blackboard space. I tried to learn from Roy's technique, but my style "jumps around" more. I too fill blackboards, even after powerpoints became the style. I am good at PowerPoint presentations for meetings, but worry that they lure students into a TV mood of "not thinking" if used excessively in class.

The three of us — Barclay, Roy and I — all stayed on as active faculty well after retirement age. We were dubbed "The Three Musketeers" by Roy. I was the first to break this tradition by retirement, but remain an honorary Musketeer!

Shortly after I started at Illinois, Art Chilton, a retired naval officer, joined the faculty to teach and do research in radiation shielding and health physics. A graduate of the Naval Academy at Annapolis, his last assignment had been as Commandant at the Naval Civil Engineering Research Lab in California. He was one of the few Naval Officers who managed to keep a beard while on active duty, showing his stature in the Navy. He became my office mate (shared faculty offices were common in those days). Art took me under his wing. He had spent more time in graduate schools (courtesy of the Navy) than I had! Thus he held two MS degrees plus a PhD in Civil and Environmental Engineering. Due to his expertise in radiation health physics, he was appointed to several key international

*Art Chilton is shown here cutting his birthday cake at a party in our house in 1959. Art taught Health Physics and Radiation Shielding at Illinois and was considered an expert on radiation detection and the effects of low levels of radiation on humans. He was one of my mentors and close friends. As a retired Naval Officer, he had many life experiences he shared with me as a young staff member. As described above, he introduced me to Gump's Department Store in San Francisco where he purchased the ceramic chicken statue shown here as a gift for my birthday.*

committees charged to set up standards. One of these committees was charged by the IAEA to investigate effects of extended exposure to low-level radiation (a very controversial topic, even today). That work consumed much of Art's time.

I learned much from Art. In classes and oral exams Art often stressed units and fundamental definitions (e.g., the subtle fundamental differences between neutron fluxes, current, fluence, and dose). I originally tended to "sluff off" on such things, thinking they were obvious. I changed my attitude when I realized that often students did not truly understand these fundamentals until "forced" to do so. Art was also a stickler for good grammar and punctuation in student papers (plus his own technical papers). I learned some grammatical rules from him about this that have stuck with me. In addition Art felt I was not very worldly, having only served briefly on active duty in the "lowly" Army (versus his beloved Navy). Thus he tried to help me (unsuccessfully) in that area also. We often travelled together to annual ANS meetings. During one in San Francisco he introduced me to his favorite store, Gumps, which handles fantastic home decorations and unique international art objects. Gumps also became a favorite of Liz and mine and we always stop by when we happen to be in San Francisco. Art knew my interest in paintings of chickens, an interest that came from my Dad raising chickens during World War II for eggs and presumably eating. (After our first butchering one for the table, my dad lost interest in doing that again, so he focused on eggs!) We had gotten to know the chickens too well, just like other "pets." While at Gumps with Art I spotted a modernistic ceramic statue of a chicken which I liked. My birthday was several weeks after we returned from the meeting in San Francisco. When I came home from work that day, there it was — the ceramic statue from Gumps with a birthday card from Art hanging on its neck! I was completely surprised and grateful to him for the thoughtfulness.

Later, when I became Chairperson of Nuclear Engineering Program, after some arm twisting, Art agreed to serve as Associate Chairperson for my first 6 years in that position. Art, with his leadership and precision gained from years in the Navy was ideal for this position. His efforts added much to the effectiveness of our department administration. Art retired from the University some years later. He had told me of his plans to combine some continued academic and research projects as Emeritus, plus relax on a farm that he and his wife Charlotte bought. It was located about 50 miles north of Champaign and they started spending weekends there. Unfortunately he became ill and passed away all too quickly before he could really enjoy his retirement.

Dan Hang was the oldest member of the faculty and he has been a "fixture" in the department over the years. Dan was a faculty member of the Electrical Engineering Department and taught fundamental circuit courses for several generations of electrical engineering students. He also liked to teach a course on Engineering Economics that he had initiated. He split time with the Nuclear Engineering Department, where he taught a course on Economics and Nuclear Fuel Management. Dan was particularly interested in the economic issues associated with fuel management. This involves such things as optimizing schedules for moving fuel elements around in a reactor and for removing them for reprocessing. He and several nuclear engineering students developed a complicated fuel management computer code that they licensed for use to several power companies with nuclear plants, including local utilities Illinois Power and Commonwealth Edison (both companies have since been purchased or absorbed by a larger company). I have always been impressed by the many company executives and presidents who remember Dan from taking his economics courses, especially those from his early teaching days. During an alumni reunion Dan is always sought out by many attendees.

I also learned much from various college administrators at the University. Bill Everitt, Dean of Engineering when I was hired, took a personal interest in me as well as other new staff. Thus he invited me to his office for discussions off and on. Dean Everitt had a laidback nature (some staff joked about entering Dean Everitt's office and finding him standing on his head while braced against a wall, presumably to relax). He was very active in major professional societies, including the American Society for Engineering Education (ASEE) which had its original headquarters at the University of Illinois (later moved to Washington, DC, as the number of members grew rapidly). Bill encouraged me to join ASEE, and I did, thinking their meetings and journal would help me improve my teaching and student advising ability. The ASEE summer meetings were generally held on university campuses. Attendees were encouraged to bring their families by offers of low price dorm rooms and a variety of family activities plus "day camps" for kids. The kids' camps were usually manned by graduate students rounded up by the host university. Thus frequently I took my family and we all enjoyed these meetings. One unplanned event became a tradition for us. Three years in a row I was paged to come out of a session because my son, Hunter (then ~3–6 years old), had been injured and taken to a hospital emergency room. One time he was hit by a metal seat on a swing at a playground. Another time he was watching a dance class for older children and stepped out on the dance floor where someone tripped over him. It came to the point that we could remember the meeting by what happened to Hunter! On one occasion Liz was in the bathroom where she heard a small girl excitedly tell her mother about a daring thing "the hunter" did that morning. Liz interrupted and said Hunter was her son. The mother was relieved to realize "the hunter" her daughter kept talking about was actually another child, not a real hunter. Dean Everitt heard about these incidents and laughed with me about them.

I had been at Illinois 3 years when Dean Everitt called me into his office and suggested that I submit a package of materials to support a proposal that I be promoted to Associate Professor with tenure. Young staff members today are extremely anxious to gain tenure. However, times were different in those days. I felt that I could easily get a job at another university or in industry if I didn't move ahead at Illinois. Thus my initial response to Dean Everitt was reserved. I said that the package would take much time to put together, and I was behind in my research. Could I wait a year? He rightly snapped back, "Nonsense — do it!" I did and was promoted.

Dean Everitt was a member of the McKinley Presbyterian Church which was on the edge of the campus and catered to students. Liz and I were members of the First Presbyterian Church in downtown Champaign. One day Dean Everitt told me that he had nominated me to serve on the McKinley session as a liaison member from First Church. I accepted and had many challenging experiences over the 6 years I served. Those were turbulent times with student rebellions occurring on campuses across the country. One by-product of this was that a few radical student groups requested space in the McKinley building for their work, which in one case included showing X-rated movies to raise money! I was strongly opposed to this, but other McKinley session members were more lenient and several such organizations, including the one showing adult movies, were permitted to use the facilities. Next McKinley became a haven for students in trouble during student "riots" at Illinois as part of the many student war protests that swept across college campuses in the United States in the late 1960s, early 1970s. Some injured student rioters were taken in and protected from police who were taking a large number of "leaders" to jail. At the time I was on leave at Cornell University so was only participating in McKinley session activities by phone and mail (see the discussion of my time at Cornell and the electron beam diode-pumped laser in Chapter 5). Just the year before, student protestors at Cornell armed with guns had forcibly taken over administrative offices in Willard Straight, the student union building. That event got front page coverage in *Time* magazine. Friends said I was crazy to go to Cornell on sabbatical in view of that. However, the following year when I was at Cornell the students had really calmed down, but student protests broke out back at Illinois. Social and political "fads" tend to start on the East and West coasts and move in to the conservative Midwest. Thus going to Cornell that year, despite the warnings from friends, worked out OK by accident! I actually ended up avoiding being in the midst of the turmoil that happened on both campuses.

While at Cornell I received the following description of events back at Illinois from friends. During the first day of the student protests, a large mob of students and some others started out marching towards the TRIGA reactor building. Fearing that mob members were out of control and might interfere with radioactive materials stored in the reactor building, the reactor staff and some police stood ready to block any attempts to enter the building. However, the mob marched right past the reactor and went on down the street to the building where a large supercomputer, called ILLIAC IV, was under construction. The students demanded it be shut down because they said it was going to be used by the Department of Defense to design weapons! University administrators had previously given news reporters a press release about the computer being sponsored by the Defense Advanced Research Projects Agency (DARPA) to do long range weather forecasting. No matter, the students claimed that this publicized purpose was a cover-up to hide actual intentions. They broke some things and blocked all entrances into the computer building for some days. Under pressure from DARPA, University officials finally moved the ILLIAC IV to NASA facilities on Mofett Field near San Jose, CA. This event is described by Wikipedia as,

> "When the computer was being built in the late 1960s, … protesters … felt that the University had sold out to a conspiracy. The protests reached a boiling point on 9 May 1970, in a day of "Illiaction". Three months after the August 24th bombing at a University of Wisconsin mathematics building, the University of Illinois decided to back out of the project, and have it moved to a more secure location. … NASA, then still cash-flush in the post-Apollo years and interested in almost anything "high tech"... formed a new Advanced Computing division, and had the machine moved to Moffett Field, CA, home of Ames Research Center."

This was an unfortunate decision for computer staff at Illinois who had a long-term reputation as leaders in design and construction of large, fast computers (other university groups generally focused on software). Thus, many of the people working on the leading commercial CRAY supercomputer had come from Illinois, and the ILLIAC was part of a famous series of advanced designs done at Illinois. The loss of ILLIAC IV set things back some years until Illinois proposed and received NSF support to create the National Center for Supercomputer Applications (NCSA). Today NCSA has NSF support to build "Blue Waters," destined to be one of the fastest supercomputers in the world. Construction is well underway for the computer and also for supplying the power source required, which equals about half again as much as used daily for the whole campus. Operation is planned to start in 2013.

When I became Chairperson of the Nuclear Engineering Program, Dan Drucker was Dean of Engineering. I served under him for about 5 years and learned much from him. Dean Drucker had tremendous talent for management and scientific research. He had come from Brown University where he chaired the theoretical mechanics department. He accepted the job with definitive, well-defined goals for increasing the national ranking of the College of Engineering. To do this he set up a number of metrics for all of the departments in the College, and he personally took much time evaluating staff and departments. In the process he also encouraged and helped departments increase their research funding levels, both from the state and the federal government. He stressed the quality of research versus quantity. Reaching into teaching practices, he encouraged graduate students to take several basic courses from departments other than their home department, selecting from required courses taken by students majoring in that department (versus "service" course versions). He also sought ways to free time for faculty to do cutting edge research. All of this worked and by the end of his Deanship, the College had gone up several steps in ratings, making Illinois the top ranked Engineering college in the Midwest.

Dean Drucker had a photographic memory. When I went to his office to discuss proposed annual salary increases for NE staff, I soon discovered he had read and remembered much about our entire faculty. He had requested that the three top (highest impact) publications in our biodatas be clearly marked. Faculty in general did not take that too seriously, thinking no one would take note or care. However, I soon found that Dean Drucker could list from the top of his head several of these publications for each faculty member. He also had an opinion about the importance of these particular publications, and he recalled other aspects of the biodata (e.g., the number of PhD and MS students supervised, honors and awards received, teaching ratings from student evaluations). Thus he had his own well-informed view about who excelled and should get the largest raises before I could even present my evaluation. When all was said and done however, my evaluations and his were not that different. In a few cases where we disagreed, I was not able to overcome his arguments and change his mind. The only time I recall really changing the Dean's mind involved an issue raised about the operational cost of the TRIGA reactor. He called me in one day to say that statistics showed the TRIGA was the most expensive laboratory in terms of dollars/student in the college. He said something had to be done about that, and I asked for time to investigate. I came back with reams of paper to prove that this was an aberration, because the real issue of how to prorate expenses between student labs, service activities such as generation of radioisotopes, research, and also training of others (e.g., operators from Illinois Power came over periodically for "refresher" training, which was beneficial because one could do many things to test skills on the TRIGA that were not possible on an operating power station or a training simulator). I claimed the benefits of the TRIGA went well beyond the student lab aspects, but some of those aspects could not be expressed numerically. In a way I managed to overwhelm the discussion with data. Dean Drucker reluctantly agreed but warned that any "missteps" relative to the TRIGA would cause another review. Fortunately that did not happen during the time he was Dean. However about 20 years later, as noted earlier, the University administration closed the TRIGA down due to liability concerns. Because I no longer had administrative responsibilities I was not directly involved in that debate. I did testify before the review committee assigned to make the decision about the importance of the TRIGA to my research.

Hiring faculty may have been a hallmark of my time as Chairperson of the old Nuclear Engineering Program. Money was reasonably available then and I managed to present strong cases for hiring faculty to the Dean and Provost. I gradually built up the largest faculty size we have had to date. Since then departures and retirements without replacements (due to budget constraints and decreasing numbers of undergraduate students as nuclear energy became unpopular after the Three Mile Island incident) have slowly cut the faculty size in half. (After some years without new hires, four young assistant professors were added in 2012, including two women.) I hired many of the present faculty in the Department. This includes Jim Stubbins, David Ruzic, and Cliff Singer. It is interesting that both David and Cliff came to Illinois from the Princeton Plasma Physics Lab to work on fusion plasmas. Their presence here greatly increased our capability and stature in the field. David still works on fusion, but has developed a large and exciting program on plasma processing. This includes research on plasma processes involved in making computer chips, and his research on this is heavily supported by major companies in the semiconductor field. Many of his recent students who have graduated have taken jobs in Silicon Valley companies involved in plasma processing. David's effort has greatly expanded the scope of our plasma work and teaching.

Cliff Singer initially had a very active research program on fusion plasmas and his students who worked with him on that have assumed important positions in the field. However he became interested in nuclear disarmament issues and eventually became the Director of the Disarmament Center on campus. This center not only covers disarmament issues but also examines many political and economic issues related to global energy structures and energy policy. Cliff stepped down as director but continues much research in these areas. This unexpected (at least to me) change in his interests has been very beneficial to NPRE, bringing in a whole new perspective on global issues related to nuclear energy.

This brings us up to the Department today. Jim Stubbins, current Head of NPRE, has done a masterful job of steering the Department through the recent stressful years when nuclear energy was out of favor and our student enrollment slowly dropped. To buffer this somewhat, Jim managed to broaden coverage of areas in NPRE and formed many international collaborations. He also managed to help faculty keep the graduate research going at a high level and has maintained high morale within the department. In 2011, with the much discussed "rebirth" of nuclear energy, the undergraduate enrollment increased from around 60 to almost 200. This presents a problem for Jim, however, because restrictions can prevent hiring of new faculty. (Fortunately funding has now increased and several new faculty members were recently hired.) In the meantime Jim used affiliated staff to good advantage to help with the increased student load on faculty. However, there are always challenges. The terrible earthquake and tsunami that struck Japan and caused the subsequent nuclear reactor problems could affect the future of the field. But the lessons learned may well make the nuclear energy option even more safe and competitive in the future.

In addition to University activities, I have always tried to stay involved in community activities that build on skills I have gained from research and teaching at the University. One activity I have found particularly fulfilling is serving as a Member of the Amateur Preacher's Association of the First Presbyterian Church in Champaign. This group fills in for ministers in small churches throughout central Illinois when they are on summer vacation. For 30 years I have given such guest sermons one or two times a summer. My sermons have often dealt with patience and optimism based on faith, similar to what I described here about my view of how to approach research. I have also discussed the issue of a possible conflict between science and religion. My belief in that the two are not in conflict because science seeks to understand the laws of nature and the universe, while faith helps us understand ethical principles to guide our choices through our life's journey.

## Further Reading

G. H. Miley, "Research and Teaching Fusion in Nuclear Engineering," *J. of Eng. Ed.*, pp. 1–10 (1971).

G. H. Miley, "Teaching and Research in Fusion Technology at U. of Illinois," *Trans. ANS.*, Vol. 16, pp. 14–15 (1973).

G. H. Miley, R. Stubbers, L. Wu, and H. Momota, "An IEC-Driven Sub-Critical Assembly for Teaching Laboratories Offering a New Generation of Sub-Critical Experiments," *14th Pacific Basin Nuclear Conf.*, Mar. (2004).

G. H. Miley, "Experience Teaching a Course on the Hydrogen Economy and Fuel Cells to Nuclear Engineers," *ANS Transactions*, Vol. 99, p. 121 (2008).

# Chapter 19

# Creation of a Small Company, NPL Associates, Inc.

*Left: Guenther Altman (right) with Heinz Hora during a dinner at my home. Guenther was one of the officers in Rockford Technologies LLC of Vancouver, Canada, that purchased license rights to IEC and nuclear reactor control methods. Rockford Technologies established a branch office in Champaign, which I headed. Rockford Technologies of Champaign was active for a half dozen years. Then, due to financial problems at the home office, it was closed and all projects were transferred back to Vancouver. I then went forward to create NPL Associates Inc. to allow competition for SBIR/STTR innovative research grants. NPL is still very active today. Right: John Sved, Daimler-Benz Aero Space, Germany, is shown outside of my house in Champaign during a visit in 1995. He was in charge of the section in Daimler-Benz Aerospace in Germany that licensed our IEC technology for use as a neutron source for neutron activation analysis (NAA). John became extremely interested and was an intense proponent of the IEC neutron source. He later left Daimler-Benz to form his own company offering IEC neutron sources for industrial NAA. I have continued IEC research at Illinois and at NPL Associates, but in recent years have concentrated more on its application to space thrusters.*

Almost all entrepreneurs think about starting a small company which becomes widely known for its innovative products and then prospers. In the 1980s I too was attracted by that thought. However, after talking to people who had done this and reading stories about early days in several extremely successful start-up companies, I changed my mind. I realized doing that would require me to leave the University and completely throw myself almost exclusively into company development. I heard horror stories about failures of such new start-ups; not only of the company, but of marriages. Statistics show that the divorce rate of founders of small start-up companies is notoriously high, regardless of whether the start-up is successful or not! The problem seems to be that many founders become overly obsessed with time-demanding company activities to the neglect of other parts of his/her life. Those were sobering thoughts. Plus, I really like university life, so I put thoughts about a start-up company aside.

However, later I realized that I could start a company and take advantage of experts and students at the University to staff it, but I could also stay at the University. I could work part-time for the company, legally, because the University generously allows a day a week for consulting. In addition I am not on contract during summer months, freeing even more time. The government Small Business Innovative Research (SBIR) program was relatively new at that time and it dawned on me that it would offer a way to accomplish my goals for the company. As noted earlier in Chapter 11, this program is designed to encourage small companies to develop innovative concepts that have commercial potential. It was set up by Congress to foster growth of new small business firms in the United States, thus growing our economy. Realizing that I could form a company to take advantage of the opportunities afforded by the SBIR program, I filed papers in Illinois to form a S-corporation named Nuclear Plasma Laboratories (NPL) Associates. I initially thought of developing Nuclear Pumped Lasers through this company, hence the name. As time passed, I simply called the company NPL Associates, Inc., or NPL for short. As discussed in Chapter 4, funding for nuclear pumped lasers essentially ended with the end of the Cold War. Thus NPL had to seek new areas for development. NPL became active fairly quickly when it received a DOE SBIR grant on fission reactor control theory with me as the PI. The proposal was based on the thought that I had developed earlier during my research days focusing on fission reactor kinetics. It was also aided by follow up work I did with Professor S. Kim, an expert in reactor control theory from Seoul University in Korea. He spent part of a sabbatical at Illinois working with me on interactive control theory. The business plan for NPL was simple, and remains my current model. I wanted to obtain money to support research (i.e., mainly support various half-time researchers committed to the project). Intellectual property would be accumulated, and sold to interested licensees. In a way, this company was like a non-profit organization, but it wasn't. My measure of success was mainly the ability to support staff, often my students, to do research that excited us and could lead to new patents.

NPL's first SBIR contact on control theory resulted in a patent and a computer code for designing the self-tuning control system described in the patent. This patent later became an important factor in my involvement in Rockford Technologies as described next. Indeed, this interaction and my involvement with Rockford Technologies is a complicated story. I will only hit a few highlights here.

At the time I founded NPL, one of my research areas was on IEC fusion devices for neutron sources (described earlier in Chapter 11). Heinz Hora, Professor of Theoretical Physics at the University of New South Wales, Australia, was my friend and collaborator. He had followed this work and had some ideas about the IEC himself. He discussed the possibility of combining some features of his concept with mine. He then filed a German patent application on that with me as co-inventor. Heinz is very active and inventive.

During this same time period he began thinking about a liquid boron control rod concept for added safety and control of fission reactors. Due to my background in fission reactor physics, he asked me for advice about several aspects of this liquid control rod concept. He eventually filed for a German patent on the control rod. I was again a co-inventor due to my contribution to the concept. I am not up on the German patent system, but Heinz could file patents easily with much less expense than needed for U.S. patents.

Heinz then came in contact with Larry Owens in Vancouver, Canada. Larry was a stock broker and also invested in start-up companies. He had set up Rockford Technologies of Vancouver through some complicated trades and was looking for a lead product. Heinz convinced him that the boron control rod could prevent future Chernobyl-type accidents and would draw funding from major nuclear reactor manufacturers around the world. In addition to new plants, Heinz argued that existing power plants might be retrofitted with this safety device. Larry was convinced and offered us stock in Rockford Technologies and his help in getting business started for the safety control rod. Thus, Heinz and I became stock holders in Rockford and began a search for "customers" for this new type of control rod. Meanwhile, Larry brought in another financial entrepreneur, John Tompkins, to help manage Rockford Technologies and get it listed on the Canadian stock exchange. About this time, I also became involved with John Sved at Daimler-Benz Aerospace, Germany. As also described in Chapter 11, John convinced his management at Daimler-Benz to set up a new group to develop the IEC neutron source for use in NAA quality control. At this point the situation got complicated. John Tompkins learned about the IEC neutron source work at Illinois and obtained a license from the University for IEC neutron source development based on our patent that had been filed through the University. Tompkins already held Hora's and my German patent rights through Rockford Technologies. He then turned around and negotiated to sell all of these license rights to Daimler-Benz. These contractual issues took time with a team of Daimler-Benz lawyers involved, but they were finally resolved. Then, as part of the agreement, my IEC group, working with John Sved and his group at Daimler-Benz Aerospace in Germany, developed a rugged version of the IEC source suitable for use in NAA in an industrial setting. This unit included simplified controls and a data collection computer setup which enabled operators with little scientific knowledge to safely and accurately run the NAA station. Several such stations were then installed in the Daimler-Benz Aerospace plant in Germany, replacing Cf-252 neutron sources that were formerly used for NAA to measure possible impurities in incoming metal ores. This replacement was desirable because Cf-252 (made by irradiation in a fission reactor) has a 2.6 year half-life, so must be replaced frequently. Due to cut-backs in its manufacture, it is becoming scarce and very expensive.

Unfortunately, a downturn in the automobile market at that time reduced money available in Daimler-Benz for new activities. As a result the original plan to market these IEC-NAA units was abandoned. The ones already constructed were used on NAA quality control systems at ore mines run by Daimler-Benz Aerospace in Germany. With the project ending, John Sved left Daimler-Benz to form his own company intended to commercialize IEC neutron sources. Daimler agreed to let him take much of the IEC testing equipment from their lab with him. Two researchers from my IEC lab who had just graduated with PhDs, Robert Stubbers and Brian Jurczyk (see the discussion of Starfire Industries in Chapter 11), worked with Sved for a short time to get things moving. I have not had contact with Sved in recent years, but he still advertises IEC neutron sources for NAA on the Internet.

John Tompkins wanted me to work on IEC outside of the University in order to protect IP and allow aggressive marketing. He happened to know about the NPL control theory patent obtained during the earlier SBIR and concluded that it had good business potential. John then convinced Larry Owens and others in Rockford Technologies to purchase the patent from NPL in exchange for stock in Rockford Technologies, and to fund a branch of Rockford Technologies in Champaign with me as its Chief Scientist. The branch was intended to seek additional U.S. government contracts. This deal was finally agreed to and signed. An office for Rockford in Champaign was opened on Springfield Avenue near the University. Lori Ballinger (now the Business Manager in Aerospace Engineering at the University) became the full-time business manager for Rockford-Champaign. We also hired one full-time staff member and several part-time consultants. However, by the time all of these contractual arrangements were accomplished, we had already done much of the IEC NAA development work at the University. John Tompkins had originally viewed that as the first project for Rockford-Champaign. Now we set out to generate new projects and funding.

One very interesting project that Rockford-Champaign soon obtained was a DOE contract to study the effect of radiation (gammas and neutrons) plus debris from ICF target implosions on reflecting and transmitting optics. This issue is termed the "last mirror problem" in ICF reactor design. The concern is that there has to be either a reflective or transmitting optic to focus the laser beam onto the target in the reaction chamber. Radiation and debris created by the target implosion can pass back out of the chamber through the same opening the beam entered, thus hitting the focusing optics. The damage to the optics and the choice of materials to reduce damage was of great interest. Rockford-Champaign hired consultants and several graduate students from the Nuclear Engineering Program and paid for testing time on the TRIGA reactor to do these studies. The TRIGA with its high neutron production during pulsing provided a unique facility for such research. The data obtained has been very important for early design studies of ICF reactors. In fact, this data was referred to in a recent elaborate study of a proposed next-step reactor (named LIFE) to follow NIF presented by LLNL staff at a recent ANS meeting.

Rockford-Champaign was quite successful with several projects lasting over 6 years. However, then the Canadian headquarters got into financial trouble and closed the Champaign branch. Later, Rockford Technologies was removed from the Canadian stock exchange and they filed for bankruptcy.

After my involvement with Rockford ended, I again turned my effort to IEC research at the University and NPL Associates. I had been continuing that work at NPL, but the Rockford-Champaign activities cut into my time. The IEC work had gradually moved into uses for space propulsion, both as a near-term electrical thruster and as a future fusion powered propulsion unit (see Chapter 14). Both thrusters share the common feature of extracting a plasma jet from the IEC for thrust. NPL has enjoyed several NASA SBIRs/STTRs on that subject, subcontracting for work in my lab in the University. These subcontract arrangements were set up with me working on only one side (either as PI for NPL or PI on the University of Illinois subcontract) to avoid any conflict of interest issues. An oversight committee was set up by the office of the Vice President for Research at the

University to periodically review activities to ensure conflicts of interest did not develop. This committee continues to oversee all contracts involving the University and NPL Associates. This arrangement has worked well and no conflicts of interests have come up.

In recent years the NPL work also moved more into the fuel cell area. Prajakti Joshi Shrestha, who worked for me in the University Fusion Studies Lab office, became business and technical coordinator for NPL. Then when her husband's job moved to the Chicago area, she set up a "second" NPL office there. NPL has been involved, either as the prime or the subcontractor, on a series of important fuel cell contracts from NASA, DARPA, Air Force, and Sandia National Labs. This has allowed NPL to gain a leading position in the special area of all-liquid fuel cells. Most recently NPL has been involved in a contract from a major U.S. company to deliver a demonstration cell to show its suitability for use in powering an unmanned underwater vehicle (UUV). Hopefully this will lead to a contract for actual units for this purpose. NPL is also working with a small company in North Carolina in a joint effort to develop ion injected IEC units as a next step towards power production and also on the $NaBH_4/H_2O_2$ Fuel Cell for near term energy storage and conversion. As discussed in Chapter 14, recent support through John Scott's office in NASA's

*Left: Prajakti Joshi Shrestha serves as business and technical manager for NPL. Among her duties, she has organized several meetings including one we hosted in 2007 at Argonne National Lab near Chicago as part of the DOE U.S.–Japan Workshop Series on IEC fusion. Joshi arranged a tour to see famous building architecture along the river in downtown Chicago. The participants, including the Japanese delegation really enjoyed the experience. Right: The next meeting was held at the University of Wisconsin, and Joshi assisted with the organization of that meeting as well. She is shown here standing on the steps of the Wisconsin capitol building.*

Johnson Space Flight Center has provided an added emphasis for the research on an injected IEC thruster called HIIPER.

In conclusion, NPL has worked out to have been more successful than I originally envisioned. Its future directions must remain fluid, however, being so strongly dependent on funding opportunities. Two directions are currently being pursued: the HIIPER thruster described in Chapter 14 and the LENR power cells discussed in Chapter 12. Relative to LENR, the joint venture setup through LENUCO noted earlier in Chapter 12 represents a new type of endeavor for NPL. It will take time to find out how this works and whether a Black Swan is involved.

## Further Reading

G. H. Miley, G. T. Park, and B. S. Kim, "Adaptive Control for a PWR using a Self-Tuning Reference Model Concept," *Proc. 1992 Top. Mtg. of Advances in Reactor Physics*, Charleston, South Carolina, March, pp. 2-129–139 (1992).

G. Miley, J. Sved, "The IEC Star-Mode Fusion Neutron Source for NAA — Status and Next-Step Designs," *Proceedings, IRRMA '99*, Raleigh, NC, Oct 3–8 (1999).

# Chapter 20

# Where Am I in the Search? What Have I Found?

*As stated at the start of this book, swans have been seen by me fairly frequently over the years, but they are not quite "black"!*

Where am I along the zigzag route to a Black Swan? As I wanted to be quantitative in finalizing these notes, I began to think in terms of numerical scores. How many mute and Black Swans have I sighted? Then, after reflecting, I began to worry about keeping score. Experiences along the path and the spirit of the goal are more important than some numerical grading. Still, I finally decided to go ahead with a graph to summarize my personal evaluation of success with sightings, which appears at the end of this chapter. Because I am naturally optimistic and enthusiastic, this graph will probably be more "positive" than the reader would conclude. Still, it is just a fun way to think about the journey.

I may have caught several fleeting "glimpses" of Black Swans. The first were early in my career involving burnable poisons and the first direct electron-beam diode pumped laser. I rate these as "near" sightings because of their significant impact on the field. However, I moved on so quickly after the fleeting "glimpse" that few associate the work with me. Thus they cannot be truly rated as a sighting. NPLs and advanced fuel fusion are different. In both I am recognized as one of the "fathers" of the field. However, this is more for cumulative early contributions to development of the field than a sudden sighting. Thus, I show a Mute Swan sighting with a dashed line to black. Low Energy Nuclear

Reactions (LENRs) are an interesting case. My transmutation results might have been a sighting, but that phenomena as a whole is still viewed with great suspicion. I feel that D-cluster reactions will surely led to a solid sighting, but much remains to be done to get into position for a clear view. In many ways, teaching and education are a solid viewing. Many students who have worked with me have already started to achieve sightings of their own. I am truly proud of that and happy for them. However, aiding others along the road to sightings is in some way different from one's own journey.

Having gone through this self-evaluation, it seems appropriate to reflect on it for lessons learned in the hope that my experiences will help someone else on their own search for a Black Swan. To start with, I am forced to question my own definition of a Black Swan event. My chart implies that I almost sighted one on several occasions, but few, besides me, realize this! I believe the Mute Swans shown in the chart are worthy of their name. However, they send out ripples in a small pond visited by scientists and users of the respective Mute Swan technology. These ripples can have many effects within this sphere of influence. However, they are not quite like the Black Swan of the financial collapse that rippled throughout the banking industry worldwide as cited by Nassim Taleb in his *Black Swan* book. Several of the near Black Swans listed (e.g., LENR power cells) could lead to such global ripples, but I have yet to achieve one of those sightings. So, have I been presumptuous to feel that I have had several sightings in the small pond sightings almost at hand? Maybe the pond has been magnified in size inside my mind. Indeed, Taleb leaves us the option of Black Swans in limited size ponds (hence not with effects that reach far beyond us and persons nearby us) when he talks about Black Swans in one's personal life. I would strongly defend my view that such searches are very important to us all. Lofty goals are good, but to avoid frustration, it is best to also have a series of more modest, possibly achievable goals in mind. That is what I did. However, I realized this best now looking back. At the time, I was too immersed in the science and technology to clearly see the full picture ahead. Then again, that is Taleb's description of a Black Swan (i.e., an event that we do not anticipate will happen until it does). Afterwards, looking back, we then say "sure, we predicted that" (especially if we think of ourselves proudly as "experts").

Next, I must ask myself why I had sightings so near at hand (e.g., burnable poisons and electron diode pumped lasers) only to walk away from the chase. In both cases, this was early in my career (the 1970s), and two things happened in each instance. First, I left KAPL and burnable poison studies for the University of Illinois. Later, I left my one year visiting appointment at Cornell University working on electron diode lasers to return to Illinois. Why didn't I realize how close I was and simply stay put? I failed to recognize the importance of those Black Swans until a year or two after the near sightings. The initial work I did in both areas got the ripples going, but they both required much continuing effort by many others to create a full ripple effect that occurred across the nuclear navy (simplified control and refueling of naval reactors) and the laser community that turned to electron-diode-pumped lasers (numerous research and commercial applications).

I might well have stayed in one of those areas, but I was already wandering, looking for other Black Swans without realizing that "one in the hand is better than two in the bush." So perhaps the lesson learned should be to stay the course, because you never know when you may enjoy a sighting. Incidentally, I only realized after the fact that I had at least initiated some ripples when some former colleagues at KAPL told me "that burnable poison stuff is taking hold and will probably change all of our future reactor designs." Likewise, a colleague at Illinois in the laser area, Joe Verdeyen, said to me shortly after I returned from the e-beam work at Cornell, "George, that electron diode pumped laser you did is great. Can you give me your copy of the paper about it? I want to build one and I know several other labs that have already started to build them." In both cases, the realization of the Black Swan that had been so close began to dawn on me. Nevertheless, I no longer was near its home pond. As time went on, I have had a few regrets about that, but fortunately, my optimistic nature rescued me and I enthusiastically continued on new searches for other ponds and swans. As I have already stated, much of the fun comes with the search.

As I have already suggested, one of my most successful searches involved teaching and education. Like many educators, I feel my duty is to lead young students into the pond where Black Swans are thought to live, and instill in them the joy of the search. I have been blessed to work with many bright, talented, and enthusiastic students over the years resulting in 56 PhDs and many MS graduates. I could describe their many accomplishments, but that would require another book, or two! Based on that, however, I rate teaching as a near sighting. I do, however, show it with a dashed line because it is, in a sense, indirect. The former students have had to take the journey themselves. I just helped them start. I say "just," but I am exceedingly proud of this accomplishment. Of all of the near sightings listed, the effects of this will ultimately generate the most numerous and largest ripples. I can enjoy seeing the ripples as I sit on the sidelines of the outer banks of their ponds.

As I was in the process of proofing this book, I became aware of a book by George E. Vaillant titled *Triumph of Experience* (Harvard Press, 2012; reviewed by Andrew Stark in the *Wall Street Journal,* November 3–4, 2012), and I believe that it is relevant to this discussion. *Triumph of Experience* is one of a series of "progress reports" on the Harvard Grant Study, which has followed over 260 graduates from Harvard over the course of seven decades. The study objective was to identify attributes that best predict a successful life in terms of achievement, income, good physical and mental health, and happy marital and parental relationships. The issues covered in the study's interviews gradually changed over the years. Initially the focus was on predicting the future of the study's subjects, but now the interviews with the surviving men, all in their 90s, generally focus on how they have "come to terms with their past". As Vaillant states it, "What a man thinks at a late stage of life much depends on how successfully he has come to terms with life's regrets." Stories taken from interviews revealed a wide array of strategies for making peace with life's missed opportunities. The final conclusion is that the future is what young men dream about. They try to shape their future, but continue to worry about the extent to which this is possible. In contrast, older men dream about past, and they try, in their minds, to reshape it. Vaillant concludes that, "For the most regret-free men in the Harvard study, the past is a work of their future."

The conclusions from this study are very relevant to my search for a Black Swan, but there is a fundamental difference which can be summarized as: "Despite your age, focus on your future rather than struggling your past." I am trying to resist the temptation to reshape the past in my mind. Thus rather than rationalizing past failures to find a Black Swan, I am continuing the search for one. And, as I already stressed, I am reveling in the search, especially if a Mute Swan is found. Perhaps this is because I am still about ten years younger than the men in the Harvard Grant Study. I hope not, and I hope I can continue looking forward, satisfied with the past, which is an exciting zigzag path even though it has not yet found the Black Swan. Indeed, by definition, a Black Swan event cannot be predicted, so may be one of some type will still happen if I am persistent! I know others who are also searching, and I hope that you, the reader, will also consider this option for your life. In this way one can be in harmony with his life's regrets without trying to reshape them.

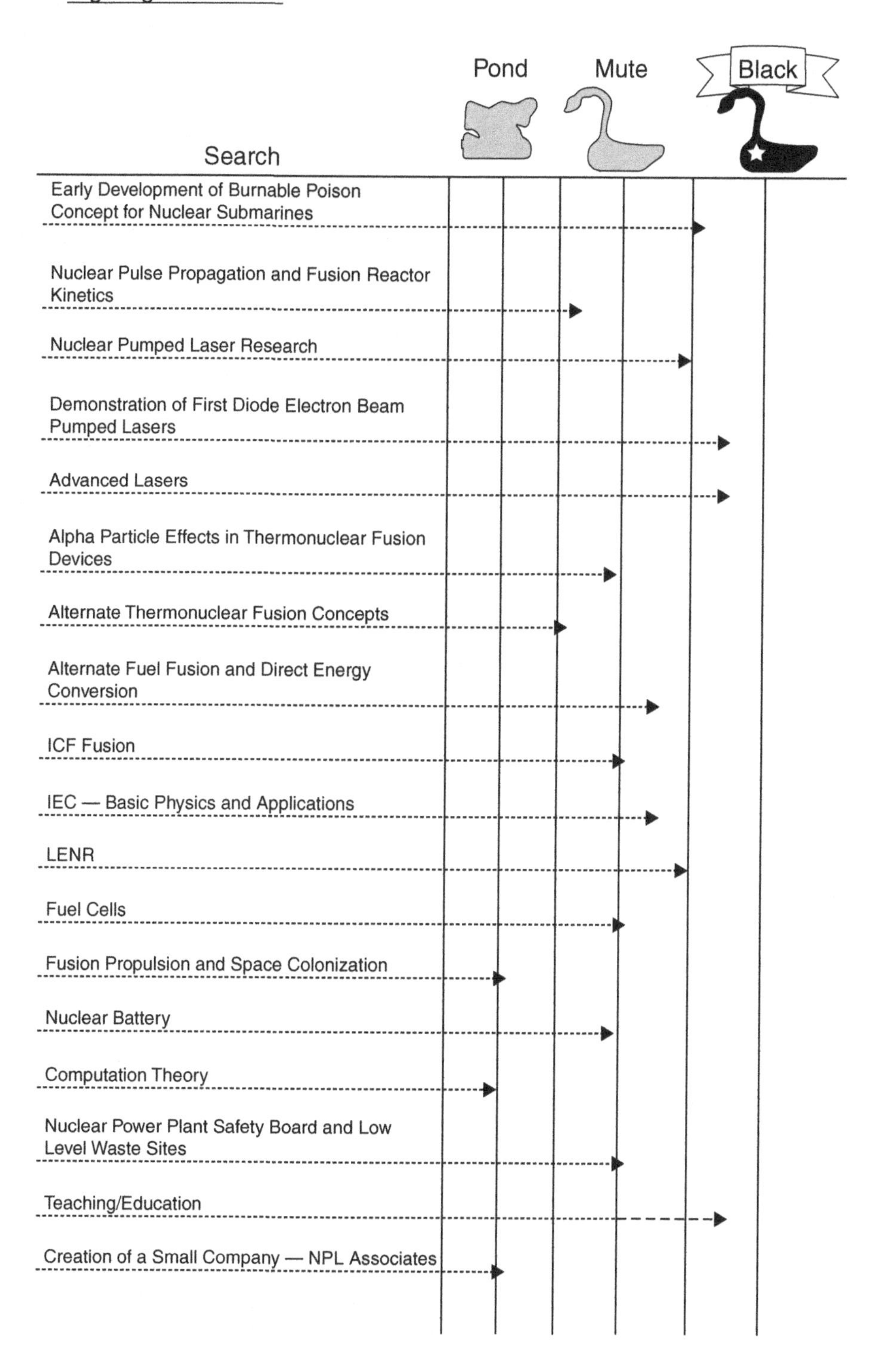
Sighting Score Card
Pond
Mute
Black
Search
Early Development of Burnable Poison Concept for Nuclear Submarines
Nuclear Pulse Propagation and Fusion Reactor Kinetics
Nuclear Pumped Laser Research
Demonstration of First Diode Electron Beam Pumped Lasers
Advanced Lasers
Alpha Particle Effects in Thermonuclear Fusion Devices
Alternate Thermonuclear Fusion Concepts
Alternate Fuel Fusion and Direct Energy Conversion
ICF Fusion
IEC — Basic Physics and Applications
LENR
Fuel Cells
Fusion Propulsion and Space Colonization
Nuclear Battery
Computation Theory
Nuclear Power Plant Safety Board and Low Level Waste Sites
Teaching/Education
Creation of a Small Company — NPL Associates

# Chapter 21

# Concluding Comments

*In the fall of 1979, during our stay in Novosibirsk, Russia (largest city in Siberia) as invited guests of the nuclear research center, Liz and I enjoyed walks in the wooded parks with beautiful colored autumn leaves. It was during one of these walks that Liz had a discussion about the Bible with the translator/guide assigned to her. The conversation was cut short, however, by someone rustling in some bushes nearby who caused the guide to fear that they were being overheard by a KGB agent (Bibles were outlawed then). That event caused a "black cloud" to hang on over the rest of the walk. As I have travelled my own zigzag path, I too have on occasion heard rustling nearby, but I have desperately tried to ignore that and focus on the joy of the path and beauty of nature. I hope you are able to do that too.*

I hope that you, the reader, have enjoyed tracing my search for the Black Swan. The Black Swan still seems to be there, but hiding among the trees and amongst the ever-present Mute Swans. We never know when a sighting may occur. Even if that wonderful event never happens, the search has been (and is still) an invigorating journey. And the Mute Swans have grown to possess a "shine" all their own. As Alexander Graham Bell once said, "When one door closes, another one opens." But if we keep pushing on the closed door we may fail to see the open one.

Are you too on a journey seeking a Black Swan? It is my hope that overlooking my personal journey through this book's reflections may give you more enthusiasm to continue. When I began, I did not think about starting the journey; rather, I just stumbled onto the path. As I went forward on the search, excitement "grew" in me about the need for new energy sources. I first verbalized this goal in the "Who's Who" biographical statement cited earlier. However, human emotions are very complicated. My drive to succeed with the journey clearly comes from a complex mixture of motivations. (Your drive may come from a different "mix" — we are all unique.) I officially started mine over 50 years ago when I received my PhD from the University of Michigan. (I attended my 50th class reunion in September 2009.) Graduation started me on the way, but it is easy to get lost unless one keeps renewing his knowledge of the countryside. Fortunately for me, that renewal comes naturally through interactions with youth (students) when one is a professor. It is hard to stand still while surrounded with students pushing you in both class and research. In addition, interactions with colleagues through meetings and research papers, both in and outside of the University, play a similar role. Thus, it is just possible that you have influenced me even though we may never have met face to face.

Another aspect of the journey that I have not stressed but seems obvious is that the path is often curvy and the surroundings full of "quicksand bogs" that can fatally trap someone who ventures too far off of the path. Plus, some wildcats may be lurking in the trees near the swan's home. We all hope to avoid these pitfalls along the way. Still, we may blunder into something that was not desired. As with the "closed door" of Alexander Bell, the best solution is to gradually work around the pitfall and look for a clear path that may be nearby but barely visible. If we keep pounding on the closed door, or keep trying to wade through the pitfall, we may fail to see the new path waiting for us. I feel I have succeeded in doing reasonably well, but in retrospect, did remain trudging through a wooded forest too long in several instances.

So, telling you about my journey has invigorated me and, I hope, encouraged you to continue on, also.

George H. Miley
January 24, 2013
Champaign, Il 61821

*Left: The search moves on with a new generation. Our oldest grandson (we have six grandchildren), Matthew Hibbs, is shown here at his graduation from high school in Bloomington, MN, in 2009 with his mother, Susan, and father, Mark, as well as his brother Michael and sister Sarah. His dad is also in a gown because he handed out diplomas representing the school board. Matthew is attending Pepperdine University (Malibu, CA), and spent an off-campus semester in Argentina where Pepperdine has a facility for that purpose. His father, Mark, thought Matthew might follow in his footsteps and study physics at Carleton College in MN. However Matthew's Black Swan is in a different direction.*

*Right: My son Hunter's wife, Lisa, is shown with their daughter, Ashley, and two sons, Jason and Ryan. It won't be long until they attend their graduations. Time flies! As I stressed earlier, we all have searches underway for Black Swans, but each swan is different. It is not so much the goal as enjoying life and helping others along the path. Enjoy yours!*

# Appendix

## Timeline of Events

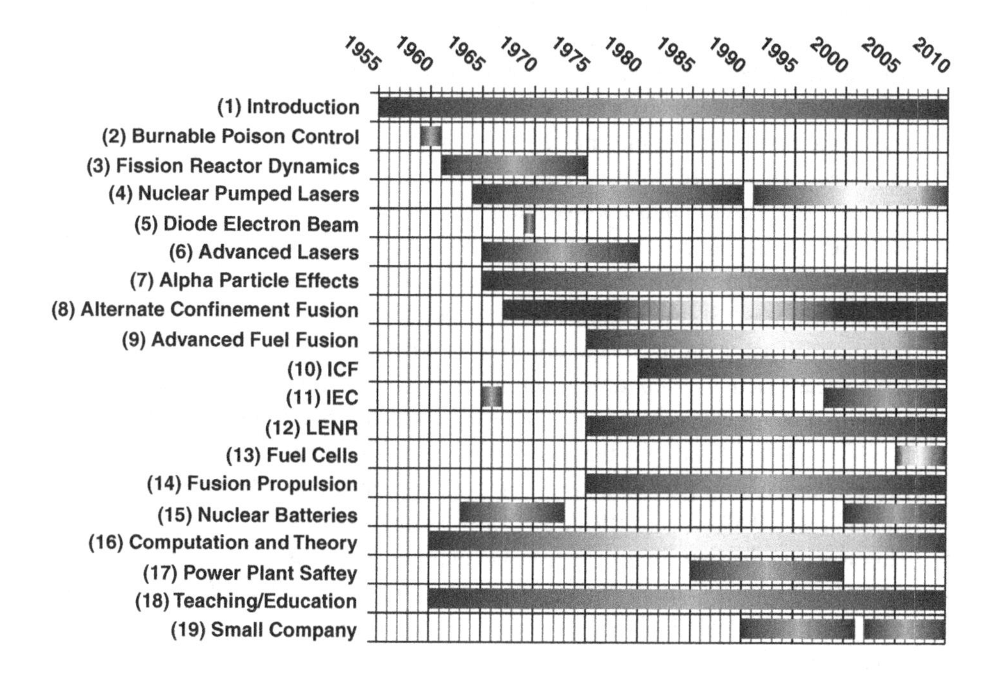


Because my research has taken a number of turns over the years (forming a "zig-zagged" path) I thought it would help to insert this chart and notes outlining this path. This can also serve as a "roadmap" to refer to when my discussions diverged or overlapped. Also, these footnotes are provided for the timeline to give a "snapshot" view of the "zigs" and "zags" along the path discussed the book.

Some of the changes in direction were evolutionary, while others resulted from a distinct event that influenced directions. The footnotes below about each zig or zag in my path correspond to the chart and to chapters in the book.

(1) Liz and I came to Illinois in the fall of 1961. The prairie scene and the advantages of this "micro-urban community" (Urbana-Champaign and the University) really grew on us. Thus, we remain here to this day. After 50 years as a faculty member, I became Emeritus in August 2011, but remain very active in both research and teaching. Our

son, Hunter, and his family live in Walnut Creek, CA, while our daughter, Susan, and her family are in Bloomington, MN. We now have six grandchildren!

(2) Upon graduation from the University of Michigan in 1959 with a PhD in Nuclear and Chemical Engineering, I joined the staff at General Electric's Knolls Atomic Power Lab (KAPL). Despite only being at KAPL 2 years (with a 6 months interruption for a tour of active duty in the Army), I was very fortunate be able to contribute to this new concept of burnable poisons. These materials are added to control rods to extend naval reactor lifetimes (time at sea before refueling). The same technique is now routinely used in commercial nuclear power plants.

(3) I came to the University of Illinois in 1961 to get in on the ground floor of the new Nuclear Engineering Program that started the previous year, and to have the opportunity to use the University's new pulsed TRIGA research reactor. This reactor was one of the first to have a pulsing capability at a university in the United States. My early research utilized TRIGA pulses for studies of nuclear reactor power plant dynamics and neutron pulse propagation.

(4) My research on Nuclear-Pumped Lasers (NPLs) came from a "brainstorm" I had in 1963 while reading one of the first books about lasers, invented only 2 years earlier. My thought was, "If lasers can be pumped by electrical excitation, why not nuclear radiation?" That question set me off on several decades of research on NPLs, which use nuclear reactions excited by neutrons to provide energy to (or "pump") the laser medium. Later, I began exchanging information with Russian colleagues working on NPLs, resulting in trips to the secret science cities in Russia. These experiences became an exciting part of that era of my research. Illinois NPL research finally ended in the 1990s as the TRIGA reactor was shut down.

(5) This work on the first direct electron-beam pumping of lasers occurred during my 1969–1970 stay at Cornell University. It grew out of my involvement in a project at Cornell's Relativistic Electron Beam Facility to develop collective acceleration of GeV ions.

(6) These studies of other advanced lasers started in the mid-1960s as a natural extension of NPL research. This work lasted about 15 years and was revived for a short time recently. Work included early research on a thermionic electron pumped laser designed to be coupled directly to fuel elements in a fission reactor as well as a recent novel electron filament Z-pinch driven concept.

(7) In the 1960s I wanted to get into fusion research in a big way but was behind because a number of well-established programs had been built up in other universities and national labs. I looked around to find some aspect that had not yet been extensively explored and decided that alpha particle effects (including heating of the plasma, losses to the wall, possible instabilities, and alpha ash buildup) represented "niche" areas where my students and I might have impact. This work has continued in various forms up to today, although there are many more competing researchers in these areas now.

(8) In the late 1960s I became interested in alternate fusion confinement in the hope of finding a way to reach practical fusion faster. That led my students and me to a series

of studies on alternate concepts, ranging from Reverse Field Pinches (RFPs) to Inertial Electrostatic Confinement (IEC), and advanced target concepts for Inertial Confinement fusion.

(9) In 1973 when I obtained a contract from ERDA (the predecessor to DOE), to write the book *Fusion Energy Conversion*, I was forced to think more deeply about how to best extract electrical energy from a fusion reactor. That led to the realization that the use of "advanced fuels" (fusion fuels such as D–$^3$He and p–$^{11}$B with high yields of charged fusion products) were essential to eventually achieve improved conversion efficiency using direct energy conversion methods. This, in turn, became a key topic in the book and led to my many years of research on both advanced fuel fusion and direct energy conversion.

(10) Research on Inertial Confinement Fusion (ICF) was a natural complement to our laser and alternate confinement fusion work at Illinois. Interest in ICF was greatly accelerated by early encounters with Professor Heinz Hora, Emeritus Head of the Theoretical Physics Department at the University of New South Wales, Australia. We first met at a 1967 ICF conference in San Diego, CA, after which he invited me to attend several workshops at the Rensselaer Polytechnic Institute (RPI) branch in Hartford, CT. Attendees included pioneers in the field from many countries, and this experience provided a real inspiration to me. These workshops ultimately became a bi-annual series on laser and particle beam interactions that Heinz Hora and I ran in Monterey, CA, for several decades. They have now grown into the current International Conference on Inertial Fusion Science and Applications (IFSA), run by the major ICF labs worldwide.

(11) My first encounter with Inertial Electrostatic Confinement (IEC) fusion in the mid-1960s resulted from my role on Bob Hirsch's thesis committee at Illinois and his subsequent pioneering work on IECs with Philo Farnsworth (inventor of the electronic television) at the ITT Farnsworth Research Laboratory in Ohio. I did some theoretical research on the topic, but soon left the field. R.W. Bussard called me in the late 1990s and asked if I would join him on a research project involving a magnetically assisted IEC concept, the "Polywell." He knew of my early IEC work and wanted to use advanced fuels in the IEC. This resulted in my starting experimental work on IEC at Illinois which continues today.

(12) Low Energy Nuclear Reactions (LENRs) are a branch of "cold fusion." I started work on cold fusion in 1975 after the dramatic public announcement by Pons and Fleischmann of "fusion in a test tube" by cold fusion. This work, now termed more accurately LENR, continues today with our concentration on use of so-called deuterium clusters formed within dislocation defects in solids. My work on nuclear transmutations created by LENR, including the fission effect in compound nuclei, has drawn considerable attention, but is only part of the objective of our research which is focused mostly on understanding of the basic physics involved.

(13) An informal discussion in 2005 about electrolysis with Nie Luo, then a post-doc researcher in my LENR lab, resulted in our current fuel cell research. The concept of an all-liquid sodium borohydride–hydrogen peroxide fuel cell came up as we glanced

through some electrochemical papers in a professional journal. We went on to obtain an international patent and various government agency grants to develop this all-liquid fuel cell. It offers an exceptionally high energy density and can operate without air because the oxygen needed comes from the peroxide. We are now considering ways to commercialize such cells through the small company, NPL Associates, Inc.

(14) Research on fusion propulsion was a natural extension of our fusion research. Thus it has been a topic of our interest since the 1960s. We used design studies to investigate use of various fusion confinement schemes including the plasma focus, field reversed devices, ICF, and IEC. In the late 1990s, George Schmidt, then a branch chief at NASA's Marshall Space Flight Center decided to expand the advanced space propulsion effort there and invited a proposal from me. I did that based on our IEC fusion experiments and corresponding propulsion design studies. Subsequently, Jon Nadler, a post doc in the IEC lab, put one of our small IEC devices in a station wagon and drove it down to Huntsville to begin work there jointly with our lab. That effort grew until a few years ago when NASA decided to cut all advanced propulsion work due to budget priorities related to the shuttle program. The area is now returning and we have submitted new proposals.

(15) Nuclear battery research was an outgrowth of insights I gained while preparing to write the ERDA (now DOE) sponsored book *Direct Conversion of Nuclear Radiation Energy* in 1967. Shortly after that the Nuclear Regulatory Commission (NRC) asked me to join a safety review board for radioisotope powered pacemakers, heavily used then for patients needing heart regulation. This was a growing area for a few years, but as chemical batteries improved and became cheaper, isotope-type pacemakers slowly faded out. I again gained interest in nuclear batteries when I was asked to join an effort by Lincoln Labs in Cambridge, MA, to join work on a DARPA project to develop nuclear batteries for use by soldiers instead of chemical batteries. That project ended after two years but the experience led to our recent new research on micro nuclear batteries.

(16) I have always enjoyed computations and simulations as a natural supplement to my experimental research. In fact, starting with my early experiences at KAPL, my philosophy has always been to combine theory simultaneously with experiments to the extent possible. This approach seems to be out of step with today's focus on specialization in the science and engineering. But, I still feel that it is an extremely effective way of doing research when working on applied science.

(17) Don Hall, a retired nuclear navy officer under Admiral Rickover, was appointed head of Illinois Power's troubled Clinton Power Station in the mid-1980s. He was brought in to apply his navy experience to the management of Clinton. Soon after arriving, he invited me to join Clinton's Nuclear Regulatory Group (NRAG), effectively an independent group charged with reviewing safety issues at the plant. Don said that he was looking for someone who was not prejudiced by prior experiences but who understood the fundamentals of nuclear power. My responsibility on NRAG was to oversee activities in Water Chemistry, Reactor Operator Training, and Emergency Preparations. During this time period I also became involved with the State of Illinois' effort to

locate a low-level nuclear waste site in southern Illinois. I headed the Governor's technical advisory committee for site evaluation. I was also appointed to the Governor's Radiation Safety Advisory Committee which oversees safety of x-ray equipment in dentist and medical facilities as well as industrial uses of radiation sources. I remain on that committee today.

(18) Teaching students through class lectures, laboratories, and supervisory research has been a big part of life since coming to the University of Illinois. I have truly enjoyed that. While I recently became Emeritus, I still continue in teaching and research as much as possible.

(19) In the 1980s, I became aware that government's Small Business Innovative Research (SBIR) program encourages proposals for innovative work in the beginning stages. In contrast, standard government agency grant programs (NSF, DOE, NASA, etc.) tend to be much more "conservative." Because I enjoy doing innovative research, I started a small company to compete for an SBIR contract. I submitted a successful SBIR proposal to develop a novel "self-tuning" control method for a nuclear power plant. Later, Rockford Technologies of Vancouver, Canada, purchased the self-tuning control programs, forming a branch of Rockford Technologies in Champaign. I served in it as the Chief Scientist. Later, after a period of some turmoil, Rockford Technologies of Champaign was dissolved. I then founded NPL Associates, Inc., which remains quite active doing energy-related research.

# Afterword

*In addition to professional interactions, Liz and I have always enjoyed being with my colleagues and students on other occasions. This is a photograph from a 2004 party at our house with faculty, grad students, and visiting international staff. We have hosted this type of event for many years. Unfortunately I do not have photographs from all such events nor space to show them if I did. These occasions and many other events must remain fond memories in our minds.*

I have greatly enjoyed working with all of my students and my many other collaborators. I have learned and grown in experience and enthusiasm from all of these interactions. So, if you read this book but don't find your name, know that I still rejoice in our association and good times together. You are certainly in my memory and hopefully I am in yours. I just need time to write another book!

Also, you will notice that these recollections are a somewhat one-sided view — usually upbeat, successful events. I certainly had a number of down periods, but my mind tends to best recall the good times. I have tried to be accurate but apologize for errors that no doubt crept in due to a fuzzy memory. As time passes, one tends to embellish past events. George Shultz, former U.S. Secretary of State, discussed this tendency in an article for the *Wall Street Journal* (December 4–5, 2010). He stated that he is well aware of how easily his memory could play tricks on him and said, "I had someone in my office scribing away constantly during discussions and meetings. These notes have enabled me to recount the accuracy of history with the richness of actual dialogue." Well, unfortunately, I don't have such notes. So as George Schultz warned, my memory may have played tricks on me at places in this book, but I do hope you understand and enjoy the spirit of these ramblings anyway.

# Index